Statistics in Theory and Practice

Statistics in Theory and Practice

Robert Lupton

PRINCETON UNIVERSITY PRESS
PRINCETON, NEW JERSEY

Published by Princeton University Press, 41 William Street,
Princeton, New Jersey 08540
In the United Kingdom: Princeton University Press, Chichester, West Sussex

Library of Congress Cataloging-in-Publication Data

Lupton, Robert, 1958-
Statistics in theory and practice / by Robert Lupton.
p. cm.
Includes bibliographical references and index.
ISBN 0-691-07429-1
1. Mathematical statistics. I. Title.
QA276.L83 1993
519.5—dc20 92-36396
CIP

This book has been composed in Lucida Bright using TEX

Princeton University Press books are printed on acid-free paper, and meet the guidelines for permanence and durability of the Committee on Production Guidelines for Book Longevity of the Council on Library Resources

http://pup.princeton.edu

Printed in the United States of America

10 9 8 7 6 5 4 3

ISBN-13: 978-0-691-07429-0 (cloth)

ISBN-10: 0-691-07429-1 (cloth)

Contents

Preface

Let me explain why I have written yet another statistics book. I use statistics extensively, and over the years I have worked quite hard to understand the theory underpinning the techniques that I use, and to understand their strengths and limitations. Books on statistics that I consulted were either "cookbooks," consisting of instructions on how to proceed when confronted with a particular problem, or else thick books full of intimidating mathematics and unfamiliar notations that required careful study to answer even quite simple questions (Kendall and Stuart, to which I frequently refer, consists of three thick volumes). This book attempts to fill a niche between these two extremes, providing an accessible but reasonably rigorous introduction to those parts of mathematical statistics that I rely on in my practical contacts with statistical problems.

The level of mathematical sophistication that I assume is rather higher than is usual in elementary texts on statistics, and is comparable to that assumed in, for example, undergraduate texts on thermodynamics. You are expected to be comfortable with calculus, and to have at least met complex analysis. I use matrices where appropriate, and employ a summation convention whenever it's convenient. A mathematician would not be happy with my proofs, but I believe that they will satisfy most scientists.

There is very little real data in this book, and very few exercises of the sort that ask you to apply the techniques presented to real problems. It was painful to exclude tables of the sort that so enliven Kendall and Stuart (e.g., the age distribution of Australian bridegrooms between 1907 and 1914, or the yields of wheat and potatoes in 48 English counties for 1936), but I felt that there were many excellent sources of such problems and that including more here would only have increased the weight and price of this book, while not providing anything not readily available elsewhere.

I have not provided tables of t, F, or any of the other common statistical distributions. One minor effect of the advent of computers has been the decline of statistical tables; it is much easier to calculate a point on a Gaussian curve than it is to find the right page in a volume of tables, and I for one find it easier to compute an incomplete gamma function than to interpolate between the values for 20 and 30 degrees of freedom in a table of χ^2. I have not spent much time discussing these numerical details; for example, you will not find a prescription for actually calculating the percentage points of the χ^2 distribution, or my thoughts on minimizing a non-linear function of 27 variables. For advice on such matters I recommend *Numerical Recipes* (Press et al. 1988). Editions are available in C, Pascal, and Fortran.

Rather than pretend that this book is based directly on original research papers, I have provided references to a small range of secondary material. References are indicated by a small superscript at the point where they are relevant, with the actual citations gathered at the end of each chapter; the abbreviations used (such as *K&S*) are explained in the section on references just before the answers to the problems. Readers who wish to explore the primary literature will find references to it in the works that I cite.

The problems are an important part of this book. They discuss details whose exposition would disturb the flow of the argument, illuminate points made in the text, and present material that doesn't quite fit anywhere else. Even if you don't solve every one of them yourself, I strongly recommend at least glancing at the answers provided.

I would like to thank the colleagues, at the University of Hawai'i and at Princeton University, who suggested that I publish this book, and the students who saw it in earlier incarnations and forced me to clarify the presentation, tighten the proofs, and fix many errors, not all of which were typographical; Michael Woodhams deserves especial thanks for helping me keep misspellings out of the completed manuscript. My thanks also to my editors at Princeton University Press; Lilya Lorrin, who displayed commendable faith in the project while it was being refereed; Sara Van Rheenen, who put up with me thereafter; and Carolyn Fox, who found an incredibly large number of missing commas. The graphs were prepared using the program SM whose authors I'd like to thank for a superior product; the first drafts were largely written in Coffee Manoa, Hawai'i, whose hospitality I gratefully acknowledge. All remaining errors are, of course, my responsibility, and I would be grateful for any comments or corrections, either sent directly to me (my internet address is rhl@astro.princeton.edu), or via P.U.P.

Statistics in Theory and Practice

1. Introduction

Statistics has a rather shady reputation of being a technique for lying with numbers; a more positive view would be that statistics is the art of deducing the properties of an infinite population given only a small sample.

For example, we are all familiar with averages, even if only in the context of cricket or baseball. If a player has figures for only 5 innings we can still calculate his batting average, but we know that it won't be very accurate. If we wait for 50 innings we believe that we will get a more reliable answer, and we feel that if we wait for 500 or 5000 innings then the value will be close to the player's *true* average, which is (alas) known only to God.

What basis is there for these beliefs? If we want to go beyond a fuzzy feeling that it is obvious that the average will behave as described we will have to consult a Statistician who will (in this case) be able to reassure us. Furthermore, we can ask Him for a range in which the *true* average lies, but here He will be a little cagey. The problem is that the player may have been incredibly lucky but may really be totally incompetent, so our Statistician will reserve the right to be wrong while analysing a certain fraction of players; if we grant Him 5% (or 10%, or 2.5%), He will be happy to give us the desired range.

If you nodded knowingly when our Statistician offered to provide a range in exchange for a percentage, you have probably been exposed to statistics before. Such a claim relies on His knowing the probability of the player being lucky, and our Friend silently assumed a particular form for this. We shall return to the question of whether such assumptions are justified.

Skeptics will doubtless have realized that after 500 innings a player's skill will probably have improved (or deteriorated), so his average over such a period will be different from that over 50 innings. Will this not invalidate our conclusions? The simple answer is Yes — such a systematic effect lies beyond the reach of statistics. Once warned of such a possibility, though, our Statistician can re-examine the data to see how fast the average is changing, and can tell us whether the change is likely to be real.

A more subtle objection is that, given that a player grows old, even God cannot know his true average as no such thing exists. The Statistician's solution is ambitious: arrange to clone a million copies of the player in question, and arrange for them to have identical childhoods, then allow them to play one game on their 21st birthdays: the average of their performances in these games is the one known to God.

Once we have been to the trouble and expense of obtaining a million copies of our player, we can allow each to follow his career, noting their performances carefully. This *population* of players can then be used to predict the probable career of our original player. We have shifted the focus from the sample to the properties of the population from which it is drawn.

Not all statisticians would accept this view of their discipline. In particular there is a growing interest in Bayesian statistics, whose adherents assign meaning to the probability of a player's having a certain level of skill. This dispute will appear in a number of contexts throughout this book, but I'm afraid that it will not be resolved.

2. Preliminaries

2.1. Basic Definitions

In statistics we are concerned with deductions about a parent population, based upon the properties of a sample. Let us consider the properties of the parent population from which our sample is drawn, and define $F(x_0)$ as being the probability that a random variable $x < x_0$; clearly $F(-\infty) = 0$ and $F(\infty) = 1$. We can then define the *probability density function* (p.d.f.) $f(x)$ by

$$f(x) = \frac{dF}{dx},$$

so

$$\Pr(x \in x, x + dx) = F(x + dx) - F(x) = f(x)\, dx.$$

(For the case of a discrete distribution, F will not be differentiable and f will consist of a collection of delta functions. Rather than worry about such matters, while deriving general results I shall assume that all distributions are continuous; if all integrals are treated as Stieltjes integrals they will degenerate into sums when required.)[1] It is obvious that

$$\int_{-\infty}^{\infty} f\, dx = 1.$$

We can easily generalize these definitions to deal with two or more random variables; for example, the probability that $x < x_0$ and $y < y_0$ is $F(x_0, y_0)$, and $f(x_0, y_0)\, dx\, dy$ is the probability that x and y lie within a volume $dx\, dy$ of (x_0, y_0). The variables are said to be *independent* if $f(x, y) = f(x)g(y)$, i.e., if the value of x has no bearing on the value of y.[2]

If we integrate over one of the variables, y say, we are left with the *marginal distribution* of x, ϕ say:[3]

$$\phi(x) = \int_{-\infty}^{\infty} f(x, y)\, dy.$$

If x and y are independent this is simply $f(x)$.

Sometimes you have the p.d.f. $g(x)$ for x, but really want it for some function $\xi(x)$. I think about this as follows:[4]

$$F = \Pr(x < x_0) = \Pr(\xi < \xi(x_0)) = \int_{-\infty}^{x_0} g(x)\, dx,$$

so the p.d.f. of ξ, $\gamma(\xi)$, is given by

$$\gamma(\xi(x_0)) = \frac{dF}{d\xi} = \frac{dF}{dx}\frac{dx}{d\xi} = g(x_0)\frac{dx_0}{d\xi};$$

i.e.,

$$\gamma(\xi) = g(x)\frac{dx}{d\xi},$$

just like changing variables in any other integral. If $\xi(x)$ is not monotonic you have to be more careful, but the basic idea remains the same.

Problem 1. Let us assume that all stars of a certain spectral type have the same rotation velocity v_r, but that the orientations of their spin axes are random. The measured rotation velocity is $v_r \sin\theta$, where θ is the angle between the line of sight and the rotation axis. What distribution does $v_r \sin\theta$ follow? What are its mean and standard deviation?

The traditional ways of describing the "centre" of a distribution include[5,6]

the *(arithmetic) mean* (μ),[7]

$$\mu = \int_{-\infty}^{\infty} x f(x)\, dx;$$

the *median*,[8]

$$1/2 = \int_{-\infty}^{\text{median}} f(x)\, dx;$$

the *mode*,

$$0 = \left.\frac{df}{dx}\right|_{\text{mode}};$$

and the *(geometric) mean* (u),[9]

$$\ln(u) = \int_{-\infty}^{\infty} \ln x\, f(x)\, dx.$$

It is found, for moderately skewed distributions, that the median is approximately $(2 \times \text{mean} + \text{mode})/3$.

Problem 2. Let us assume that a p.d.f. is approximately $N(0,1)$ (see section 3.1) and write

$$f(x) = \Phi(x)\left(1 + \alpha(x^3 - 3x)\right),$$

where Φ is the usual Gaussian p.d.f. $\Phi(x) = 1/\sqrt{2\pi}\exp(-x^2/2)$. Calculate f's mean, median, and mode on the assumption that α is small, and show that in the same approximation[10]

$$\frac{\text{median} - \text{mode}}{\text{mean} - \text{mode}} = \frac{2}{3}.$$

(This is an example of a *Gram-Charlier series.*[11] The polynomial $x^3 - 3x$ is a Hermite polynomial that is chosen to give f zero mean and unit variance; the skewness may be shown to be 6α.)

The "width" of a distribution may be described by one or more of the following: the *variance* (or, equivalently, the *standard deviation* σ),[12,13,14]

$$V(x) \equiv \sigma^2 = \int_{-\infty}^{\infty} (x - \mu)^2 f(x)\,dx;$$

the *absolute deviation* (about d),[15]

$$\delta = \int_{-\infty}^{\infty} |x - d| f(x)\,dx;$$

and the *mean deviation* δ,

$$\delta = \int_{-\infty}^{\infty} |x - \mu| f(x)\,dx.$$

The absolute deviation is minimized if taken about the median.

Problem 3. Prove that the absolute deviation is minimized about the median.

These integrals may not converge. For example, the *Cauchy distribution* (section 3.7) doesn't have a mean deviation, let alone a variance. The next two paragraphs introduce measures that are guaranteed to exist.

The *full width at half maximum* (FWHM) is defined as the distance between the two points where $f(x)$ first falls to half of its maximum value. If the distribution has more than one maximum we can still define a FWHM, but it may not mean much.

Problem 4. What is the FWHM of the Gaussian distribution

$$f(x) = \frac{1}{\sqrt{2\pi}\sigma} e^{-(x-\mu)^2/2\sigma^2}?$$

There are also various *interquantile* ranges, where quantiles are defined analogously to the median, i.e., as the values of x where the cumulative distribution reaches some specified value. Some of these quantiles have special names, for example, the the interquartile range is the difference between the upper and lower quartiles where[16]

$$\frac{3/4}{1/4} = \int_{-\infty}^{\frac{\text{upper}}{\text{lower}}\text{ quartile}} f(x)\,dx$$

and the semi-interquartile range is half the interquartile range.

A convenient notation is to write an *expectation value* (i.e., the "average value") as

$$\langle y \rangle = \int_{-\infty}^{\infty} y\, f(x)\, dx,$$

so (e.g.) $\sigma^2 = \langle (x-\mu)^2 \rangle$. The expectation value of y is also sometimes written as $E(y)$.

The mean and variance are easily generalized to higher orders, and the higher moments of a distribution are defined as[17]

$$\begin{aligned} \mu'_r &= \langle x^r \rangle \\ \mu_r &= \langle (x-\mu'_1)^r \rangle. \end{aligned}$$

The mean is $\mu = \mu'_1$, and the variance $\sigma^2 = \mu_2$. Note that

$$\begin{aligned} \mu_2 &= \langle (x-\mu)^2 \rangle \\ &= \langle x^2 - 2\mu x + \mu^2 \rangle \\ &= \langle x^2 \rangle - 2\mu \langle x \rangle + \mu^2 \\ &= \langle x^2 \rangle - \mu^2 \end{aligned}$$

(when I write an unqualified μ it's the mean). Some quantities based on the third and fourth moments have special names, $\mu_3/\mu_2^{3/2}$ being called the *skewness*,[18] and $\mu_4/\mu_2^2 - 3$ being called the *kurtosis*[19] (the 3 results in a Gaussian distribution having zero kurtosis).

It is also possible to define mixed moments, of which the most important is the *covariance*[20]

$$\sigma_{xy} = \langle (x - \langle x \rangle)(y - \langle y \rangle) \rangle$$

of x and y. If the variables are independent the covariance vanishes, although variables can be dependent even if $\sigma_{xy} = 0$. An example is two variables x and y whose joint p.d.f. consists of the filled circle $x^2 + y^2 < 1$. The covariance is obviously zero, but the range of x depends on the value of y, and vice versa. The covariance of two variables depends on the units in which they are measured, so it is sometimes convenient to use the *product-moment correlation coefficient* instead,[21,22,23] also known as *Pearson's correlation coefficient*,

$$\rho_{xy} = \frac{\sigma_{xy}}{\sigma_x \sigma_y},$$

which always has a value in the range $[-1, 1]$.

2.2. The Distribution of $g(x, y)$ Given the Distribution of x and y

A simple way to find the approximate distribution of $g(x)$ is to use a heavily truncated Taylor series:[24,25]

$$g(x) = g(\mu) + \left.\frac{dg}{dx}\right|_{\mu} (x - \mu),$$

where $\mu = \langle x \rangle$, so

$$\langle g \rangle = g(\mu)$$

and

$$\left\langle (g - \langle g \rangle)^2 \right\rangle = \left.\frac{dg}{dx}\right|_{\mu}^{2} \sigma_x^2.$$

This can of course be generalized to functions of many random variables, which need not be independent; e.g., if g is a function of two variables x and y,[26]

$$\sigma_g^2 = \left.\frac{\partial g}{\partial x}\right|_{\mu_x}^{2} \sigma_x^2 + 2 \left.\frac{\partial g}{\partial x}\right|_{\mu_x} \left.\frac{\partial g}{\partial y}\right|_{\mu_y} \sigma_{xy} + \left.\frac{\partial g}{\partial y}\right|_{\mu_y}^{2} \sigma_y^2,$$

where σ_{xy} is, as before, the covariance.

This expansion is only exact if the function g is linear in all of the variables. In general a more careful analysis is required, and it is in any case required if we want to know the distribution of g (for the linear case the distribution is easily found using characteristic functions if the variables are independent, a technique that we will come to shortly). The argument is very similar to that used to change variables;[27] as an example consider the distribution of $g(x,y) = xy$ and assume that x and y are independent, so the cumulative probability distribution of g is given by

$$F_g = \int_{-\infty}^{0} f_X(x)\,dx \int_{g/x}^{\infty} f_Y(y)\,dy + \int_{0}^{\infty} f_X(x)\,dx \int_{-\infty}^{g/x} f_Y(y)\,dy$$

(draw a picture if you are confused; lines of constant g are rectangular hyperbolas). By differentiating with respect to g and noting that $f_g = dF_g/dg = 1/x\,dF_g/d(g/x)$,

$$f_g = -\int_{-\infty}^{0} f_X(x) f_Y(g/x)\frac{dx}{x} + \int_{0}^{\infty} f_X(x) f_Y(g/x)\frac{dx}{x};$$

if f_X and f_Y are symmetrical about the origin, this can be reduced to

$$f_g = 2\int_{0}^{\infty} f_X(x) f_Y(g/x)\frac{dx}{x}.$$

2.3. Characteristic Functions

Let us consider[28]

$$\begin{aligned}\phi(t) &= \left\langle e^{ixt}\right\rangle \\ &= \int_{-\infty}^{\infty} e^{ixt} f(x)\,dx \\ &= 1 + (it)\mu_1' + \frac{(it)^2}{2!}\mu_2' + \cdots.\end{aligned}$$

ϕ is simply the (inverse-)Fourier transform of f so[29]

$$f(x) = \frac{1}{2\pi}\int_{-\infty}^{\infty} e^{-ixt}\phi(t)\,dt.$$

ϕ is called the *characteristic function* of f, and has a number of nice properties:

1. It can easily be used to derive the moments of a distribution,[30] either by inspecting the expansion of ϕ in powers of it, or by noting that

$$\mu_r' = \left.\frac{d^r \phi}{d(it)^r}\right|_{t=0}.$$

(If you are wondering why there's an i in the definition of ϕ, and apparently in every formula that involves it, it's there in order to make the integrals converge and to make the inversion formula simple.)

2. Because of the inversion formula there's a one-to-one mapping between f and ϕ, so if I can prove a result in terms of characteristic functions it must be true in terms of the corresponding p.d.f.'s too, and often it's much easier to deal with ϕ.

3. Let $z = x + y$, and assume that x and y are independent. In that case[31]

$$\phi_z(t) = \left\langle e^{it(x+y)} \right\rangle = \left\langle e^{itx} \right\rangle \left\langle e^{ity} \right\rangle = \phi_x(t)\phi_y(t),$$

so it's easy to obtain the distribution of z. If you want to do it directly,

$$f(z) = \int_{-\infty}^{\infty} f_x(u) f_y(z-u)\, du,$$

a result that is either obvious or else follows from the convolution theorem of Fourier analysis. Note that if we had had the sum of n distributions the characteristic function approach becomes no more difficult, while the direct approach involves an $(n-1)$-dimensional integral.

4. If $z = x/n$, then

$$\phi_z(t) = \left\langle e^{itx/n} \right\rangle = \phi(t/n).$$

5. If $z = (x - \mu)/\sigma$, then

$$\phi_z(t) = e^{-it\mu/\sigma}\phi(t/\sigma).$$

6. Even if you can't invert ϕ to get f, you can always get the moments of the distribution directly from ϕ. In fact the inversion can often be done, as the inversion integrals are well suited to the powerful technique of contour integration in the complex plane (see section 4.1 for an example).

We can find the distribution of a sum of n independent variables x_i in a more direct way, which may perhaps help to explain the motivation for introducing characteristic functions in the first place:

$$f(z) = \int_{\sum x_i = z} f(x_1) \cdots f(x_n)\, dx_1 \cdots dx_n,$$

which is hard to evaluate due to the linear constraint imposed on the x_i. Let us move the problem under the integral sign:

$$f(z) = \int f(x_1) \cdots f(x_n) \delta(\Sigma x_i - z)\, dx_1 \cdots dx_n,$$

where there is now no restriction on the domain of integration. We know that

$$\delta(x) = \frac{1}{2\pi} \int_{-\infty}^{\infty} e^{-itx}\, dt,$$

(a result which is the basis of Fourier analysis), so we can write

$$f(z) = \frac{1}{2\pi} \int_{-\infty}^{\infty} e^{-itz} f(x_1) e^{itx_1} \cdots f(x_n) e^{itx_n}\, dx_1 \cdots dx_n.$$

This is now a repeated integral, so writing

$$\phi(t) = \int_{-\infty}^{\infty} f(x) e^{itx}\, dx$$

we see that

$$f(z) = \frac{1}{2\pi} \int_{-\infty}^{\infty} e^{-itz} \phi^n(t)\, dt,$$

as shown above.

References

1: *K&S* 1.23
2: *K&S* 1.34
3: *K&S* 1.33
4: *K&S* 1.35
5: *Bevington* 2.1
6: *Bevington II* 1.4
7: *K&S* 2.3
8: *K&S* 2.8
9: *K&S* 2.5
10: *K&S* 2.11
11: *K&S* 6.17
12: *K&S* 2.19
13: *Bevington* 2.1
14: *Bevington II* 1.4
15: *K&S* 2.18
16: *K&S* 2.14
17: *K&S* 3.1
18: *K&S* 3.31
19: *K&S* 3.32
20: *K&S* 3.27
21: *K&S* Ex. 10.6, 26.9
22: *Bevington* 7.1
23: *Bevington II* 11.2
24: *Bevington* 4.1
25: *Bevington II* 3.2
26: *K&S* 10.6
27: *K&S* 1.26, 11.9
28: *K&S* 3.5, 4.1
29: *K&S* 4.3
30: *K&S* 3.5
31: *K&S* 4.16

3. Some Common Probability Distributions

3.1. Gaussian Distribution

For reasons that should become clear presently a p.d.f. of the following form is of great importance in statistics:[1,2,3]

$$dF(x) = \frac{1}{\sqrt{2\pi}\sigma} e^{-\frac{(x-\mu)^2}{2\sigma^2}}\, dx.$$

This is the famous *normal* or *Gaussian distribution*, often written as $N(\mu, \sigma^2)$.

Its characteristic function is given by

$$\begin{aligned}
\phi(t) &= \frac{1}{\sqrt{2\pi}\sigma}\int_{-\infty}^{\infty} e^{-\frac{(x-\mu)^2}{2\sigma^2}+itx}\, dx \\
&= \frac{1}{\sqrt{2\pi}\sigma}\int_{-\infty}^{\infty} e^{-\frac{1}{2\sigma^2}((x-\mu)^2-2it\sigma^2(x-\mu))+it\mu}\, dx \\
&= \frac{1}{\sqrt{2\pi}\sigma} e^{it\mu}\int_{-\infty}^{\infty} e^{-\frac{1}{2\sigma^2}(x'-it\sigma^2)^2-\frac{t^2\sigma^2}{2}}\, dx' \\
&= e^{it\mu-t^2\sigma^2/2}.
\end{aligned}$$

This is the standard way of dealing with such integrals.

It's easy to see that $\mu_1' = \mu$, $\mu_2 = \sigma^2$, $\mu_{2n+1} = 0$, and $\mu_4 = 3\sigma^4$, so both the skewness and kurtosis of a Gaussian are zero.

Unfortunately the indefinite integral of f can't be done analytically, but it is widely tabulated or you can express it in terms of one of the error functions erf or erfc.[4]

3.2. Multivariate Gaussian Distributions

Sometimes we are interested in n variables, each of which individually follows a Gaussian distribution.[5] In general they will not all be independent, so the p.d.f. is not simply the product of the usual Gaussian distributions; instead, it is given by

$$f(\mathbf{x}) = \frac{1}{(2\pi)^{n/2}|V|^{1/2}} \exp(-(\mathbf{x}-\boldsymbol{\mu})^T V^{-1}(\mathbf{x}-\boldsymbol{\mu})/2),$$

where V is called the *covariance matrix* and is symmetric and positive definite. Its determinant is written as $|V|$, and it is easy to see that its elements are given by

$$V_{ij} = \left\langle (x_i - \bar{x}_i)(x_j - \bar{x}_j)\right\rangle,$$

so the leading diagonal consists of the variances, and the other terms the covariances, of the variables. The distribution of $\mathbf{x}$ is called a *multivariate normal* or *multivariate Gaussian distribution.*

We can change variables to diagonalize V, in which case the p.d.f. reduces to the product of n independent Gaussian variables (this is, indeed, the easiest way to see that the elements of V^{-1} are truly the [co]variances).

> **Problem 5.** If the marginal distributions of a two-dimension p.d.f. $f(x, y)$ are both Gaussian, is f a bivariate Gaussian?

3.3. Log-Normal Distribution

If x follows an $N(0, 1)$ distribution, and $x = \gamma + \delta \ln y$, then y is said to follow a *log-normal distribution.*[6] The p.d.f. is given by

$$dF = \frac{\delta}{\sqrt{2\pi}} e^{-(\gamma+\delta \ln y)^2/2} \frac{dy}{y}$$

and is plotted in the accompanying figure for a number of values of γ and δ. If we had started with an $N(\mu, \sigma^2)$ distribution we would still have had only two free parameters (not four, as you might naïvely have thought). It is, however, possible to bodily shift the curves to the right or left, i.e., to set $x = \gamma + \delta \ln(y - \alpha)$, in which case $f(y)$ is only non-zero in the range $[\alpha, \infty]$.

3.4. Poisson Distribution

Let us consider an event such as breaking a plate, which occurs with a constant but small probability, and ask what is the probability of breaking $0, 1, 2, \ldots$ on a given day.[7] If the probability of breaking one in time t is t/τ, the probability of having broken exactly n at time $t + dt$ is

$$p_n(t + dt) = p_{n-1}(t)\frac{dt}{\tau} + p_n(t)\left(1 - \frac{dt}{\tau}\right),$$

so

$$\frac{dp_n}{dt} = \frac{1}{\tau}(p_{n-1} - p_n)$$

($p_{-1} = 0$). Multiplying by the integrating factor $\exp(t/\tau)$, this can be written as

$$\frac{d}{dt}\left(p_n e^{t/\tau}\right) = \frac{1}{\tau} p_{n-1} e^{t/\tau}$$

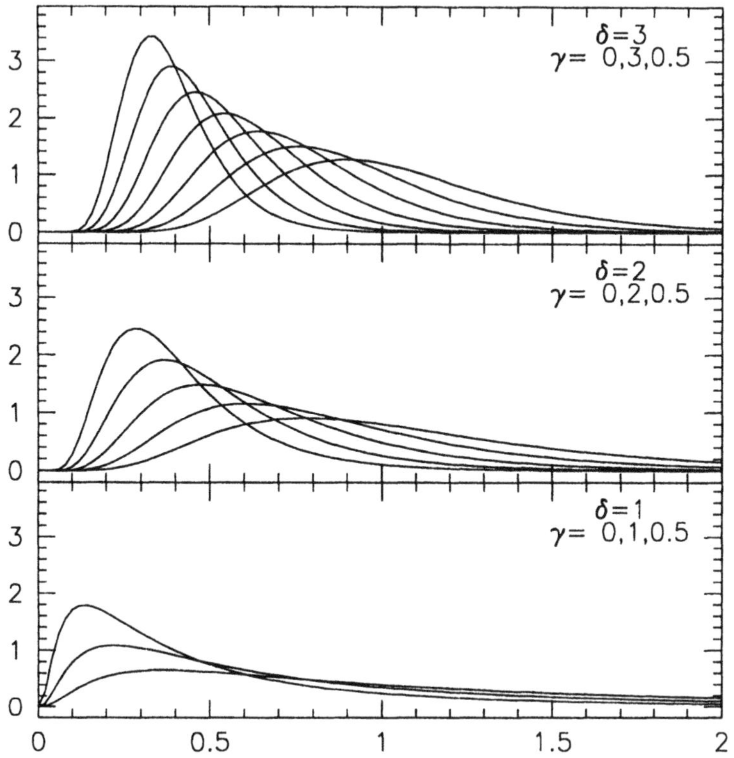

The log-normal distribution for various values of γ and δ.

or (writing $\pi_n \equiv p_n e^{t/\tau}$)

$$\frac{d\pi_n}{d(t/\tau)} = \pi_{n-1},$$

whose solution is obviously

$$\pi_n = \frac{(t/\tau)^n}{n!}\pi_0.$$

The equation for p_0 is simply

$$\frac{dp_0}{dt} = -\frac{1}{\tau}p_0,$$

so $p_0 = \exp(-t/\tau)$ and

$$p_n(t) = \frac{(t/\tau)^n}{n!}e^{-t/\tau};$$

or, writing $\mu = t/\tau$ (we shall soon see that this is indeed the mean),

$$p_n = \frac{\mu^n}{n!}e^{-\mu},$$

which is the Poisson distribution.[8,9]

Its characteristic function is[10]

$$\begin{aligned}\phi &= \sum_{n=0}^{\infty} \frac{e^{itn}\mu^n}{n!}e^{-\mu} \\ &= e^{\mu(e^{it}-1)}.\end{aligned}$$

Differentiating a couple of times, and setting $t = 0$, shows that $\mu_1' = \mu$, $\mu_2' = \mu^2 + \mu$, and $\mu_2 = \mu$, so the mean and the variance of the distribution are both equal to μ.

If we reduce the distribution to normalized form by making the substitution $n' = (n-\mu)/\sigma = (n-\mu)/\sqrt{\mu}$, the characteristic function becomes

$$\phi = \exp\left(\mu\left(e^{it/\sigma} - 1 - it/\sigma\right)\right),$$

and, as $\mu \to \infty$, this becomes

$$\phi = \exp\left(\mu\left(\frac{(it)^2}{2\sigma^2} + O(1/\sigma^3)\right)\right).$$

We know that $\sigma^2 = \mu$, so

$$\phi = \exp\left(\frac{-t^2}{2}\right)$$

to order $1/\sqrt{n}$. This is the characteristic function of a Gaussian with mean zero and standard deviation one, so we deduce that the large μ limit of a Poisson distribution is a Gaussian, with mean μ and standard deviation $\sqrt{\mu}$.

Problem 6. If I have two independent Poisson variables, x_1 and x_2, with means μ and $n\mu$, what is the distribution of $x_1 + x_2$? (Do the calculation directly, without using characteristic functions.) What is the distribution of $\sum_{i=0}^{n} x_i$, where all of the x_i have the same mean, μ? Use characteristic functions to confirm your result.

Problem 7. Show that

$$\sum_{r=0}^{n} \frac{\mu^r e^{-\mu}}{r!} = \frac{1}{n!}\int_{\mu}^{\infty} t^n e^{-t}\, dt,$$

which is an incomplete gamma function, and as such is readily evaluated.

Problem 8. Most astronomers use neither photographic plates nor their eyes to study the heavens, but devices such as CCDs that produce digital signals proportional to the number of photons that fall on the detector. If I detect n photons, what is the statistical uncertainty in n? (*Hint:* the arrivals of photons are independent events.)

The exponential distribution is closely related to the Poisson, as you will see when you solve the next problem.

Problem 9. If the probability that a friend's cat will want to be let out in a time interval dt is dt/τ, what is the probability that she'll have t minutes of peace?

3.5. Binomial Distribution

Before discussing the *binomial distribution*, I shall prove a lemma: Consider

$$\Lambda(n) = \left(1 + \frac{a}{\sqrt{n}} + \frac{b}{n}\right)^n.$$

If n is large, then

$$\begin{aligned}\ln \Lambda &\sim n\left(\frac{a}{\sqrt{n}} + \frac{b}{n} - \frac{1}{2}\left(\frac{a}{\sqrt{n}} + \frac{b}{n}\right)^2\right)\\ &\sim a\sqrt{n} + b - a^2/2 + O(1/\sqrt{n})\end{aligned}$$

and

$$\lim_{n\to\infty} \Lambda = e^{a\sqrt{n}+b-a^2/2} \quad \blacksquare$$

If I have some process with only two outcomes, A and B, with respective probabilities p and q ($p + q = 1$), and I carry out the process n times, the probability of r A's and $n - r$ B's is well known to be[11,12,13]

$$P(r) = \binom{n}{r} p^r q^{n-r},$$

where

$$\binom{n}{r} \equiv {}^nC_r \equiv \frac{n!}{r!(n-r)!}$$

is the number of ways of choosing r A's out of n letters (A's and B's). The binomial theorem assures us that

$$\sum_{r=0}^{n} \binom{n}{r} p^r q^{n-r} = (p+q)^n,$$

so the total probability is, reassuringly, unity.

Problem 10. Show that

$$P(r) = \binom{n}{r} p^r q^{n-r}.$$

The characteristic function for a binomial process is[14]

$$\phi(t) = \left\langle e^{irt} \right\rangle = \sum_{r=0}^{n} \binom{n}{r} p^r e^{irt} q^{n-r} = (pe^{it} + q)^n = (1 + p(e^{it} - 1))^n.$$

We can calculate the moments of the distribution easily enough:

$$\frac{d\phi}{d(it)} = npe^{it}(pe^{it} + q)^{n-1}$$
$$\frac{d^2\phi}{d(it)^2} = n(n-1)p^2e^{it}(pe^{it} + q)^{n-2} + npe^{it}(pe^{it} + q)^{n-1};$$

putting $t = 0$ in these expressions gives

$$\mu_1' = np,$$
$$\mu_2' = n^2p^2 + npq,$$

and

$$\mu_2 = npq.$$

Problem 11. Show directly (i.e., without using characteristic functions) that a binomial has mean np and variance npq.

Problem 12. A dog walking down a street stops every time that it passes a tree, and being too stupid to remember which way it was going, it has an equal probability of continuing in the same direction or of turning back. After visiting n equally spaced trees, what are its mean and r.m.s. (*root mean square,* $\langle d^2 \rangle^{1/2}$) distances from its starting point? This problem is an example of a class of problems known as random walks.

Let us now consider the limit where $n \to \infty$ and $p \to 0$ in such a way that $\mu = np$ remains constant. We can write the characteristic function as[15]

$$\phi = \left(1 + \frac{\mu}{n}(e^{it} - 1)\right)^n ,$$

and so, using my lemma with $a = 0$,

$$\phi \to e^{\mu(e^{it}-1)}$$

which is the characteristic function of a Poisson distribution with mean μ: the limit of a binomial, as the number of trials becomes infinite but the mean number of events remains constant, is a Poisson distribution.

An alternative limiting case is when $n \to \infty$ while p and q remain constant.[16] In this case both the mean and variance also become infinite, so we must normalize the resulting distribution. Changing variables to $r' = (r - \mu)/\sigma$ we see that

$$\phi' = e^{-it\mu/\sigma}\left(pe^{it/\sigma} + q\right)^n .$$

If we now remember that $\mu = np$ (we'll substitute for σ later),

$$\begin{aligned} \phi' &= \left(e^{-itp/\sigma}pe^{it/\sigma} + qe^{-itp/\sigma}\right)^n \\ &= \left(pe^{itq/\sigma} + qe^{-itp/\sigma}\right)^n . \end{aligned}$$

As $n \to \infty$, $\sigma = \sqrt{npq}$ also gets very large so we can expand the exponentials, giving

$$\begin{aligned} \phi' &= \left(p\left(1 + \frac{it}{\sigma}q - \frac{t^2}{2\sigma^2}q^2\right) + q\left(1 - \frac{it}{\sigma}p - \frac{t^2}{2\sigma^2}p^2\right) + O(\sigma^{-3})\right)^n \\ &\sim \left(1 - \frac{pqt^2}{2\sigma^2}(p+q)\right)^n \end{aligned}$$

or, substituting for σ^2 and recollecting that $p + q = 1$,

$$\phi' = \left(1 - \frac{t^2}{2}\frac{1}{n}\right)^n$$
$$\sim e^{-t^2/2}.$$

The binomial distribution tends to a Gaussian $N(np, npq)$ as n becomes large.

Problem 13. In a certain region of Canada there are, on average, 2 moose per lake. What distribution would you expect the number of moose (meese?) per lake to follow? If I find 5 moose on a lake, what is the probability that this would have arisen by chance? What is the probability of finding 5 or more? If you approximate the distribution by a Gaussian, what is the probability of 5 or more? What would your reaction be if I told you that I'd visited 19 other lakes before coming to this one?

Problem 14. In preparation for the general election I bought 20 hyacinths, expecting half to be red (Labour) and half to be blue (Conservative); when they flowered I discovered that 16 were blue and I immediately suspected a plot. The Conservative candidate assured me that the standard deviation of the number of blue flowers was 4, and therefore that the discrepancy wasn't significant. Should I vote for him?

Problem 15. I have a large number of pens on my desk, of which a fraction q won't write. If I want to find n that work, what is the probability that I will first try r bad ones? Find r's first and second moments and note that $\mu_1' < \mu_2$, in contrast with the Poisson ($\mu_1' = \mu_2$) and the binomial ($\mu_1' > \mu_2$). Show that if μ_1' is fixed but $n \to \infty$, the distribution of r becomes Poisson. (The distribution of r is an example of a *negative binomial*.)[17]

3.6. Multinomial Distribution

A natural generalization of the binomial distribution is the *multinomial*, where instead of two possible outcomes of an experiment there are N, each occurring with probability p_i. If the ith outcome happens n_i times,

$$P(n_1, \cdots, n_i, \cdots, n_N) = \frac{n!}{n_1! \cdots n_i! \cdots n_N!} p_1^{n_1} \cdots p_i^{n_i} \cdots p_N^{n_N},$$

where $\sum n_i = n$. The first two moments are easily shown to be[18]

$$\langle n_i \rangle = np_i$$

and

$$\left\langle n_i n_j \right\rangle = n(n-1)p_i p_j + np_i \delta_{ij}.$$

Problem 16. Calculate the characteristic function for the multinomial, and find the first two moments.

3.7. Cauchy Distribution

The p.d.f.

$$dF = \frac{1}{\pi} \frac{1}{1+(x-\mu)^2}\, dx$$

is known as the *Cauchy* or *Lorentzian distribution.*[19,20,21] It's easy enough to calculate the characteristic function directly (use contour integration and note that there are poles as $x = \mu \pm i$),[22] but it's even easier to know the answer and do the inversion integral instead. Consider $\phi = e^{it\mu - |t|}$, whose p.d.f. is given by

$$\begin{aligned} f(x) &= \frac{1}{2\pi} \int_{-\infty}^{\infty} e^{-|t|} e^{-i(x-\mu)t}\, dt \\ &= \frac{1}{2\pi} \int_{0}^{\infty} e^{-t(1+i(x-\mu))}\, dt + \text{c.c.} \end{aligned}$$

(c.c. is the complex conjugate)

$$\begin{aligned} &= \frac{1}{2\pi} \frac{1}{1+i(x-\mu)} + \text{c.c.} \\ &= \frac{1}{\pi} \frac{1}{1+(x-\mu)^2}, \end{aligned}$$

and therefore the characteristic function of the Cauchy distribution is $\exp(it\mu - |t|)$. Because ϕ isn't differentiable at $x = \mu$, no series expansion in powers of it exists, and the Cauchy distribution has no moments. If you think of the mean as being the "middle" of a distribution, you can think of μ as being the mean of a Cauchy distribution, although doing so would arouse the ire of a professional statistician. This lack of moments is also obvious from the form for $f(x)$, as the integrals fail to converge.

Problem 17. If x and y are independent $N(0,1)$ variables, show that the distribution of $z = x/y$ is a Cauchy distribution with $\mu = 0$.[23] (*Hint:* setting up the calculation is similar to the procedure for finding the t-distribution in section 6.2, but you must be careful as y can be negative, unlike s.)

Problem 18. During the revolution one of my comrades was executed by the president's guard. I know that she was placed against an (infinite) wall, that the firing squad was placed one quetzetl from the wall, and that the soldiers shot in random directions (but that all their rifles were horizontal). What is the distribution of bullet holes in the wall? How should I find the best place to lay my wreath of red roses to the martyr's memory?

3.8. Beta Distribution

Another distribution that is sometimes useful is the *beta distribution*,[24] which has two parameters α and β and whose p.d.f. is given by

$$dF(x;\alpha,\beta) = \begin{cases} \frac{1}{B(\alpha,\beta)} x^{\alpha-1}(1-x)^{\beta-1}\,dx & \text{if } 0 \le x \le 1, \\ 0, & \text{otherwise} \end{cases}$$

($B(\alpha,\beta)$ is a complete beta function; hence the name of the distribution). Its mean and variance are

$$\mu_1' = \frac{\alpha}{\alpha+\beta}$$

and

$$\mu_2 = \frac{\alpha\beta}{(\alpha+\beta)^2(\alpha+\beta+1)}.$$

The beta distribution is closely related to the binomial distribution, but it is included here as a flexible functional form to describe almost any unimodal distribution in $[0,1]$, as may be seen in the accompanying figure.

If $\alpha = \beta = 1$, the beta distribution's p.d.f. becomes a constant and the distribution is known as the *rectangular* or *uniform distribution.*

Problem 19. What are the mean and variance of

$$y = \int_{-\infty}^{x} f(x)\,dx,$$

where f is x's probability density function?

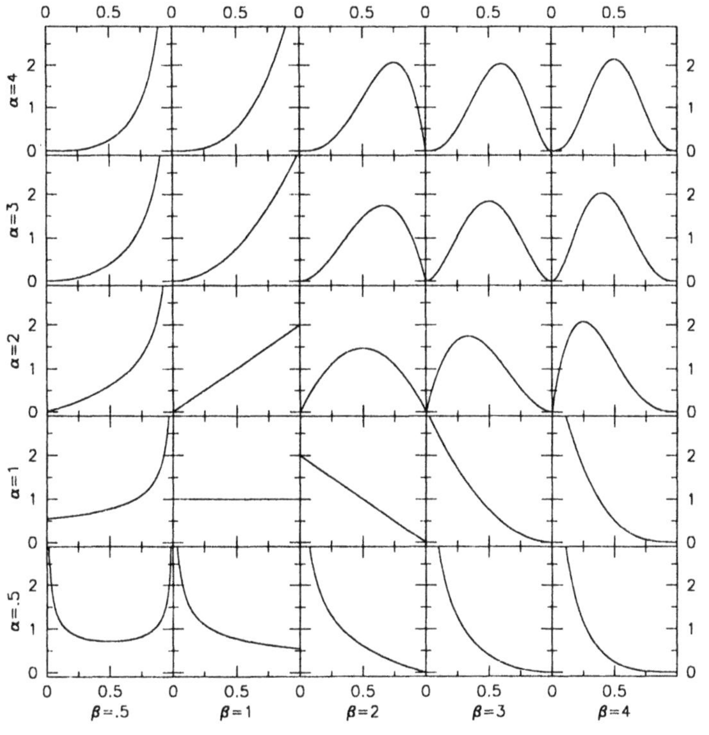

The beta distribution for a number of values of α and β.

References

1: *K&S* 5.20

2: *Bevington* 3.3

3: *Bevington II* 2.3

4: *NR* 6.2

5: *K&S* 15.1, 16.23

6: *K&S* 6.29

7: *K&S* 5.8

8: *Bevington* 3.2

9: *Bevington II* 2.2

10: *K&S* 5.8

11: *K&S* 5.2

12: *Bevington* 3.1

13: *Bevington II* 2.1

14: *K&S* 5.4

15: *K&S* 5.8

16: *K&S* Ex. 4.6

17: *K&S* 5.14

18: *K&S* 5.30

19: *K&S* Ex. 3.3

20: *Bevington* 3.4

21: *Bevington II* 2.4

22: *K&S* Ex. 3.13

23: *K&S* Ex. 11.21

24: *K&S* 6.6

4. Distributions Related to the Gaussian

Largely because of the central limit theorem, the Gaussian distribution plays a central rôle in statistics, so much so that a number of related distributions are also sufficiently important to warrant discussion.

4.1. χ^2 Distribution

The distribution of the sum of the squares of independent $N(0,1)$ variates is called the χ^2 distribution.[1] It is possible to calculate its p.d.f. directly, or to exploit what we know about characteristic functions.

If x is an $N(0,1)$ variate it's easy enough to find x^2's characteristic function:

$$\begin{aligned}\phi_{x^2} &= \left\langle e^{itx^2}\right\rangle \\ &= \frac{1}{\sqrt{2\pi}}\int_{-\infty}^{\infty} e^{itx^2-x^2/2}\,dx \\ &= \frac{1}{(1-2it)^{1/2}},\end{aligned}$$

so the characteristic function of $X^2 \equiv \sum_1^n x_i^2$ is $(\phi_{x^2}(t))^n$, or[2]

$$\phi_{X^2} = \frac{1}{(1-2it)^{n/2}}.$$

Problem 20. Show that

$$\int_{-\infty}^{\infty} e^{-itz}(1-it)^{-p}\,dt = 2\pi e^{-z}\frac{z^{p-1}}{(p-1)!}.$$

(*Hint:* use contour integration.)

We can find X^2's p.d.f. by inverting ϕ_{X^2}, so writing

$$f(X^2) = \frac{1}{2\pi}\int_{-\infty}^{\infty} \phi_{X^2}(t)e^{-itX^2}\,dt$$

and using the result that you proved in the last problem, we see that

$$dF = \frac{1}{2^{n/2}(n/2-1)!}e^{-X^2/2}X^{n-2}\,dX^2.$$

If you want this in terms of X rather than X^2, note that $dX^2 = 2X\,dX$, so

$$dF = \frac{1}{2^{n/2-1}(n/2-1)!}e^{-X^2/2}X^{n-1}\,dX.$$

This is the χ^2 distribution with *n-degrees of freedom*; X^2 is called a χ^2_n variable.

The integral of a χ^2 distribution can be evaluated, it's an incomplete gamma function, and it is both widely tabulated and easily computed.[3]

> **Problem 21.** Show that a χ^2_n distribution has mean n and variance $2n$. Show also that the χ^2 distribution is asymptotically Gaussian. (*Hint:* standardize (i.e., $X^2 \to (X^2 - \mu)/\sigma$), and remember the lemma at the top of section 3.5).

If X^2_m is the sum of squares of m independent $N(0,1)$ variates, we know that it follows a χ^2_m distribution; similarly, X^2_n follows a χ^2_n distribution, so $X^2_m + X^2_n$ (being the sum of $m+n$ independent $N(0,1)$ variates) is a χ^2_{m+n} variate. Conversely, if $Y^2_n + Y^2_m$ follows a χ^2_{n+m} distribution, Y^2_n follows a χ^2_n distribution, and Y^2_m and Y^2_n are independent, then Y^2_m is a χ^2_m variate, a result that we shall find useful.

4.2. χ^2 Distribution: Linear Constraints

If all the n elements x_i of the vector $\mathbf{x}$ are independent $N(0,1)$ variables we now know that $\mathbf{x}.\mathbf{x} \equiv \mathbf{x}^T\mathbf{x}$ has a χ^2_n distribution ($\mathbf{x}^T$ is $\mathbf{x}$'s transpose).

Let us next consider the more general case of the distribution of[4]

$$X^2 = \mathbf{x}^T A \mathbf{x},$$

where A is some symmetric matrix. We know that A possesses a complete set of eigenvectors $\ell^{(i)}$ and associated eigenvalues $\lambda^{(i)}$. It is convenient to employ the summation convention; for the purposes of this section let us agree only to sum over repeated Greek indices. Our eigenvectors then have the familiar properties that

$$A_{\alpha\beta}\ell^{(i)}_\beta = \lambda^{(i)}\ell^{(i)}_\alpha$$

and

$$\ell^{(i)}_\alpha \ell^{(j)}_\alpha = \delta_{ij},$$

(if some of the $\lambda^{(i)}$ are degenerate we can still construct such a set by Gram-Schmidt orthogonalization). We may expand $\mathbf{x}$ in terms of our $\boldsymbol{\ell}^{(i)}$: $\mathbf{x} = \sum_i a_i \boldsymbol{\ell}^{(i)}$, where the a_i are given by

$$a_i = \mathbf{x}.\boldsymbol{\ell}^{(i)} = x_\alpha \ell_\alpha^{(i)}.$$

The a_i, being linear combinations of $N(0,1)$ variates, are themselves Gaussian. We can calculate

$$\begin{aligned}\langle a_i \rangle &= \langle x_\alpha \ell_\alpha^{(i)} \rangle \\ &= \langle x_\alpha \rangle \ell_\alpha^{(i)} \\ &= 0\end{aligned}$$

and

$$\begin{aligned}\langle a_i a_j \rangle &= \langle x_\alpha \ell_\alpha^{(i)} x_\beta \ell_\beta^{(j)} \rangle \\ &= \langle x_\alpha x_\beta \rangle \ell_\alpha^{(i)} \ell_\beta^{(j)} \\ &= \delta_{\alpha\beta} \ell_\alpha^{(i)} \ell_\beta^{(j)} \\ &= \ell_\alpha^{(i)} \ell_\alpha^{(j)} \\ &= \delta_{ij}\end{aligned}$$

showing that the a_i are independent $N(0,1)$ variates (we have shown that they are uncorrelated, and this implies independence for Gaussian variables).

We can calculate X^2 in terms of the a_i:

$$\begin{aligned}X^2 &= A_{\alpha\beta} x_\alpha x_\beta \\ &= \sum_{ij} a_i a_j A_{\alpha\beta} \ell_\alpha^{(i)} \ell_\beta^{(j)} \\ &= \sum_{ij} a_i a_j \lambda^{(i)} \ell_\beta^{(i)} \ell_\beta^{(j)} \\ &= \sum_{ij} a_i a_j \lambda^{(i)} \delta_{ij} \\ &= \sum_i a_i^2 \lambda^{(i)}.\end{aligned}$$

If the $\lambda^{(i)}$'s are all equal (or zero) this is a multiple of a χ^2 variable.

Now consider the case where the matrix A is idempotent, i.e., $A^2 = A$. It's clear that this means that $\lambda^{(i)2} = \lambda^{(i)}$, so each $\lambda^{(i)}$ must be either 0 or 1; let the last k of them be 0. The number of non-zero eigenvalues (in this case $n - k$) is the *rank* of the matrix. Our expression for X^2 now becomes

$$X^2 = \sum_{i=1}^{n-k} a_i^2,$$

so X^2 follows a χ^2_{n-k} distribution.[5] The k vanishing eigenvalues correspond to k linear constraints upon the x_α, which is why the distribution is said to have $n - k$ degrees of freedom. Note that the rank of an idempotent matrix is equal to its trace.

Problem 22. Consider a set of n variables ϵ_i that follow a multivariate Gaussian distribution (section 3.2) with covariance matrix V. Show that $\boldsymbol{\epsilon}^T V^{-1}\boldsymbol{\epsilon}$ is a χ^2_n variate. (*Hint:* you can find variables $\mathbf{x} = R\boldsymbol{\epsilon}$ with $\langle x_i x_j \rangle = \delta_{ij}$.)

It is always possible to write a symmetric matrix A in the form $A = RR^T$, so we can write $X^2 = \mathbf{x}^T RR^T\mathbf{x} \equiv (R^T\mathbf{x})^T(R^T\mathbf{x})$. Now consider a second quadratic form $Y^2 = \mathbf{x}^T B\mathbf{x} \equiv \mathbf{x}SS^T\mathbf{x}$. X^2 and Y^2 are independent if $\langle (R^T\mathbf{x})(S^T\mathbf{x})^T \rangle = 0$, i.e., if $R\langle \mathbf{x}\mathbf{x}^T \rangle S^T = RS^T = 0$, which is true if (and only if) $AB \equiv RR^TSS^T = 0$. This result is known as *Craig's theorem.*[6]

Problem 23. Show that a symmetric matrix A can always be factorized as $A = RR^T$.

Problem 24. Generalize Craig's theorem to the quadratic forms $X^2 = \boldsymbol{\epsilon}^T A\boldsymbol{\epsilon}$ and $Y^2 = \boldsymbol{\epsilon}^T B\boldsymbol{\epsilon}$, where $\boldsymbol{\epsilon}$ is a Gaussian variable with zero mean and covariance matrix V.

4.3. Student's *t*-Distribution

Another important distribution related to the Gaussian has p.d.f.[7]

$$dF_t = \frac{(\nu/2 - 1/2)!}{(\nu\pi)^{1/2}(\nu/2 - 1)!}\frac{1}{(1 + t^2/\nu)^{(\nu+1)/2}}\,dt$$

and is known as *Student's t-distribution* with ν degrees of freedom.[†] It is easy to show that $\mu = 0$ and $\sigma^2 = \nu/(\nu - 2)$. As $n \to \infty$ the t-distribution

[†] "Student" is the pseudonym of W. S. Gosset, who encountered a variety of statistical problems while working as a brewer for Guinness.

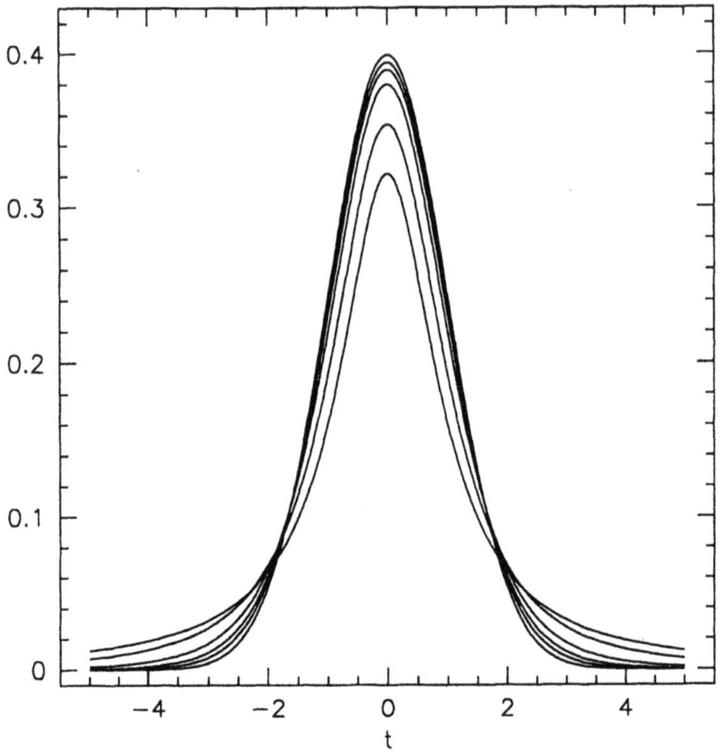

The p.d.f. of Student's t-distribution for 1, 2, 5, 10, and 20 degrees of freedom; also plotted is a Gaussian, the large n limit of the t-distribution.

becomes an $N(0,1)$ Gaussian; for $\nu = 1$ the t-distribution is a Cauchy distribution.

I shall postpone a discussion of the t-distribution's origin until we have discussed sampling, which is the context in which Student's t arises (see section 6.2). The cumulative p.d.f. is tabulated, or it can be calculated as an incomplete beta function.[8]

4.4. *F*-distribution

We are quite often interested in the ratio[9]

$$F_{\alpha\beta} = \frac{\chi_\alpha^2/\alpha}{\chi_\beta^2/\beta},$$

which is known as *(Fisher's) F* or the *variance-ratio distribution.* $F_{\alpha\beta}$'s distribution function is

$$dF_F = \frac{\alpha^{\alpha/2}\beta^{\beta/2}F^{\alpha/2-1}}{B(\alpha/2, \beta/2)\,(\alpha F + \beta)^{(\alpha+\beta)/2}}\,dF,$$

where $B(\lambda, \mu)$ is a complete beta function. When β is large enough for the appropriate moments to exist we find that

$$\mu_1' = \frac{\beta}{\beta - 2}$$

and

$$\mu_2 = \frac{2\beta^2(\alpha + \beta - 2)}{\alpha(\beta - 2)^2(\beta - 4)}.$$

It is sometimes useful to notice that the distribution of $F_{1,\beta}$ is the same as the distribution of t_β^2.

The ratio, f, of variances estimated from two Gaussian populations follows the F-distribution, and we can use the departure of f from unity to test whether the two distributions have equal variance.

Tables of F appear in almost all books on statistics, or it can be readily calculated.[10] Because F depends on 2 parameters, full tables would be 3-dimensional, so tables of F usually just give the critical values of the distribution; for example, the 95% and 99% points.

References

1: *K&S* 11.2, 11.5, 16.2
2: *K&S* 16.3, Ex. 3.6
3: *NR* 6.2
4: *K&S* 15.11
5: *K&S* 15.11
6: *K&S* 15.13
7: *K&S* Ex. 11.8, 16.10
8: *NR* 6.3
9: *K&S* Ex. 11.20, 16.15
10: *NR* 13.4

5. Sampling

We have been considering p.d.f.'s; that is, concentrating upon the properties of the population. In fact experiments only yield numbers with each number drawn (we imagine) from a certain p.d.f.: instead of $f(x)$ we measure $\{x_1, x_2, \ldots, x_n\}$. How should we proceed if we want to learn something about f?

If we are content with only knowing a few parameters of f, such as the mean, the median, or the variance, we can calculate the corresponding quantities for the x_i and hope (or show) that they are good estimates of the parent's parameters. It is also possible to place constraints on the form of f using techniques that we shall discuss in Chapter 14.

5.1. Estimating μ and σ

Let us first try to estimate the population mean, and start by calculating the sample mean [1,2,3]

$$\bar{x} = \frac{1}{n}\sum_{i=1}^{n} x_i.$$

(Roman quantities traditionally refer to sample properties, while Greek symbols are used for population properties.) Is $\bar{x}$ a good estimate of μ? This is a question with several answers, the first of which is given by calculating $\langle \bar{x} \rangle$:

$$\begin{aligned}
\langle \bar{x} \rangle &= \frac{1}{n}\left\langle \sum x_i \right\rangle \\
&= \frac{1}{n}\sum \langle x_i \rangle \\
&= \langle x \rangle \\
&= \mu.
\end{aligned}$$

We say that $\bar{x}$ is an *unbiased estimator* of μ, which simply means that its expectation value equals the parameter that it is supposed to estimate.[4] Note that not all estimators are unbiased — consider using the median as an estimator of the mean of a skewed distribution such as a log-normal.

Problem 25. One of your colleagues has built a machine to measures values of a variable t; t is known to follow an exponential distribution with (unknown) mean τ (see problem 9).

If he has measured n values of t, how would you advise him to find an unbiased estimate of τ? If he now confesses that his machine is incapable of detecting values greater than T, how would you change your advice? If it then transpires that one of his students had added a gadget to measure all values of t, but that none larger than T were actually seen, how would your advice change? Finally, how would you react if it turned out that the student's gadget didn't work?[5]

The next answer to our question concerns how accurate an estimate $\bar{x}$ is of μ, which is a question about $\bar{x}$'s probability distribution.†In general this is not the same as f, the p.d.f. of the x_i's parent population. We can calculate the required distribution using what we remember about characteristic functions. Using properties 3 and 4 from section 2.3 we see that

$$\phi_{\bar{x}}(t) = (\phi_x(t/n))^n .$$

Sometimes we may be prepared to settle for the variance of $\bar{x}$ (that is, the variance of $\bar{x}$'s p.d.f.), which is easily calculated by differentiating with respect to it:

$$\begin{aligned}
\phi_{\bar{x}} &= \phi^n \\
\phi'_{\bar{x}} &= \phi^{(n-1)}\phi' \\
\phi''_{\bar{x}} &= \frac{n-1}{n}\phi^{(n-1)}\phi'^2 + \frac{1}{n}\phi^{(n-1)}\phi'',
\end{aligned}$$

or, setting $t = 0$ and remembering that $\phi(0) = 1$, $\phi'(0) = \mu'_1$, $\phi''(0) = \mu'_2$,

$$\begin{aligned}
\phi_{\bar{x}}(0) &= 1 \\
\phi'_{\bar{x}}(0) &= \mu'_1 \\
\phi''_{\bar{x}}(0) &= \frac{n-1}{n}\mu'^2_1 + \frac{1}{n}\mu'_2,
\end{aligned}$$

so

$$\mu_{\bar{x}} = \mu$$

†What do I mean by "$\bar{x}$'s probability distribution"? Think for a moment of the million clones created in the introduction, for each of whom (which?) I can calculate $\bar{x}$, and I can then use these million values to estimate $\bar{x}$'s p.d.f.: it is a well-defined concept even though I really only have one sample.

and

$$\sigma_{\bar{x}}^2 = \frac{1}{n}\left(\mu_2' - \mu_1'^2\right) = \frac{\sigma^2}{n}.$$

The variance of $\bar{x}$ is smaller than the variance of x by a factor of n, and its square root is sometimes called the *standard error of the mean.*[6] Because the value of $\bar{x} \to \mu$ as $n \to \infty$, $\bar{x}$ is called a *consistent estimator*[7] of μ.[†]

This factor of $1/n$ is so ubiquitous that it is perhaps worth looking at the distribution of the mean from a slightly different perspective. Consider $x = \mu + \delta$, where δ is a random variable with equal probability of being $\pm\Delta$. The mean of n values of x is $\bar{x} = \mu + 1/n \sum \delta_i$, and the distribution of $\bar{x}$ depends on the behavior of $\sum \delta_i$, i.e., on a random walk (cf. problem 12); $\sum \delta_i$ has variance $n\Delta^2$ and therefore $\bar{x}$ itself has standard deviation $\Delta/\sqrt{n}$.

Problem 26. Show directly (i.e., without using characteristic functions) that the variance of $\bar{x}$ is σ^2/n, independent of the distribution of x.

Now consider the population's variance. The obvious estimate is the sample variance

$$s^2 = \frac{1}{n}\sum(x - \bar{x})^2,$$

but is this unbiased?[8]

$$\begin{aligned} s^2 &= \frac{1}{n}\sum(x - \bar{x})^2 \\ &= \frac{1}{n}\sum x^2 - \frac{2}{n}\bar{x}\sum x + \frac{1}{n}\bar{x}^2\sum 1 \\ &= \frac{1}{n}\sum x^2 - \bar{x}^2, \end{aligned}$$

so

$$\begin{aligned} \left\langle s^2 \right\rangle &= \frac{1}{n}\sum\left\langle x^2\right\rangle - \left\langle \bar{x}^2\right\rangle \\ &= \left\langle x^2\right\rangle - \left\langle \bar{x}^2\right\rangle \\ &= (\sigma^2 + \mu^2) - (\sigma^2/n + \mu^2) \\ &= \frac{n-1}{n}\sigma^2 \end{aligned}$$

[†]Formally, an estimator t for a parameter θ is consistent if, given $\epsilon > 0$, $\exists\, n_0$ s.t. $\Pr(|t - \theta| < \epsilon)\ \forall\, n > n_0$.

(we've just shown that $V(\bar{x}) = \sigma^2/n$). We see that s^2 is biased, although $ns^2/(n-1)$ isn't. In this case our naïve faith that all we have to do to estimate properties of the population is to calculate the corresponding properties of the sample has let us down; the problem being that $\bar{x}$ and the x_i aren't independent, as $\bar{x}$ is estimated from the sample.

A similar (but more tedious) calculation to that which gave us the variance of $\bar{x}$ shows that[9]

$$\sigma_{s^2}^2 = \frac{n-1}{n^3}\left((n-1)\mu_4 - (n-3)\mu_2^2\right)$$
$$\sim \frac{1}{n}\left(\mu_4 - \mu_2^2\right).$$

As $n \to \infty$, $\sigma_{s^2}^2 \to 0$, so s^2 is biased but consistent. For a Gaussian parent the exact result becomes

$$\sigma_{s^2}^2 = \frac{2(n-1)}{n^2}\sigma^4.$$

If you are interested in the distribution of the standard deviation rather than the variance, the simplest way to proceed is to use Taylor's theorem as in section 2.2:

$$V(s) = \left(\frac{1}{2s}\right)^2 V(s^2).$$

For a Gaussian, and if $n \gg 1$, this becomes $s^2/(2n)$. Problem 35 treats the Gaussian case exactly.

It is also possible to have unbiased but inconsistent estimators (e.g., $\bar{x}$ as an estimator of the central point of a Cauchy distribution, as you will see when you solve problem 29).

5.2. Efficiency of Estimators

We know that $\bar{x}$ is an unbiased estimator of the population mean μ with variance σ^2/n, but we might reasonably ask if we couldn't achieve a lower variance by using some other statistic; the sample median perhaps, or maybe the mode.

For the case of a Gaussian parent we shall show, while discussing the properties of maximum likelihood estimators in section 10.1, that no unbiased estimator of the population mean has a smaller variance than $\bar{x}$ (it is said to attain the *minimum variance bound*, the MVB). When faced with real data, which come with no comfortable guarantee that their parent is Gaussian, we may still be tempted to use a statistic (such as the median) that is less sensitive to extreme points in the data. We shall soon show that for a Gaussian population the variance of the sample median is

$$\frac{\pi}{2}\frac{\sigma^2}{n};$$

i.e., its standard deviation is about 1.253 times larger than that of the sample mean. The *efficiency* of an estimator is defined in terms of how large a sample would be required to obtain a given accuracy:[10] relative to the mean the median has (for a Gaussian population) an efficiency of $2/\pi = 64\%$. Because in this case the mean attains the MVB, we may simply say that the median has an efficiency of 64%.

If the parent population is not Gaussian the median may be very much better than this. For example, for a Cauchy distribution the sample mean has zero efficiency relative to the median (see problem 29).

Another example comes from the use of the sample mean deviation $d \equiv 1/n \sum |x_i - \bar{x}|$ to estimate the population variance. It's easy to show that

$$\delta = \sqrt{\frac{2}{\pi}}\sigma$$

for a Gaussian, so we can estimate the standard deviation as

$$\sqrt{\frac{\pi}{2}}d,$$

but what is the efficiency of this procedure?

For large n, d's variance is[11]

$$V(d) = \frac{\sigma^2 - \delta^2}{n}.$$

If we assume a Gaussian parent the sample standard deviation s has variance $\sigma^2/2n$, and d's variance becomes $\sigma^2(1 - 2/\pi)/n$; the variance of $\sqrt{\pi/2}\,d$ is therefore $\sigma^2(\pi - 2)/2n$ and the efficiency (relative to s) is

$$\frac{1}{\pi - 2} = 88\%.$$

Problem 27. Show that

$$\left\langle V\left(\frac{1}{n}\sum |x_i - \mu|\right)\right\rangle = \frac{1}{n}\left(\sigma^2 - \delta^2\right) = \frac{\sigma^2}{n}\left(1 - \frac{2}{\pi}\right),$$

where the second equality holds for a Gaussian, and argue that, for large n, this formula also gives the variance of $(1/n)\sum |x_i - \bar{x}|$.

Even though we cannot find an estimator of μ with variance smaller than the MVB, we may still be able to find a *biased* estimator t, which is better in the sense that it has smaller mean-square error:[12]

$$\left\langle (t-\mu)^2\right\rangle < \frac{\sigma^2}{n}.$$

Problem 28. If $t = a\bar{x}$, find the value of a that gives the minimum mean-square error, and show that it is less than the MVB. If $\mu = \sigma$ is the difference likely to be important?

References

1: *K&S* 10.3
2: *Bevington* 2.2
3: *Bevington II* 4.1
4: *K&S* 9.24, 17.9
5: *K&S$_V$* 31.7
6: *K&S* 9.23, 10.4
7: *K&S* 17.7
8: *K&S* 10.4, 12.3, 17.9
9: *K&S* 10.4, 12.5
10: *K&S* 17.29
11: *K&S* 10.13
12: *K&S* 17.30

6. Distributions of Sample Statistics

We have calculated the first two moments of the p.d.f. of $\bar{x}$, but what about the distribution itself? Let us first consider the case where $x = N(\mu, \sigma^2)$:

$$\phi_x(t) = e^{it\mu - t^2\sigma^2/2}$$

and, as usual,

$$\begin{aligned}\phi_{\bar{x}} &= \phi_x^n(t/n) \\ &= e^{it\mu - t^2\sigma^2/2n},\end{aligned}$$

so the distribution of $\bar{x}$ is $N(\mu, \sigma^2/n)$. We already knew the values of the mean and variance, but we didn't know that the distribution would be Gaussian.

Now consider a general distribution for which the moments exist, and normalize using the substitution $\bar{x}' = (\bar{x} - \mu)/(\sigma/\sqrt{n})$, so the distribution of $\bar{x}'$ has zero mean and unit variance. The corresponding characteristic function (cf. section 2.3) is

$$\begin{aligned}\phi_{\bar{x}'} &= \left(e^{-\mu it/(\sigma\sqrt{n})}\phi_x(t/(\sigma\sqrt{n}))\right)^n \\ &= e^{-\mu it\sqrt{n}/\sigma}\phi_x^n(t/(\sigma\sqrt{n})).\end{aligned}$$

For large n, we can expand the characteristic function in powers of t, giving

$$\phi_{\bar{x}'} = e^{-\mu it\sqrt{n}/\sigma}\left(1 + \frac{it\mu_1'}{\sigma\sqrt{n}} - \frac{t^2\mu_2'}{2\sigma^2 n} + O(n^{-3/2})\right)^n,$$

which, as $n \to \infty$, becomes (using the lemma proved at the top of section 3.5)

$$\begin{aligned}\phi_{\bar{x}'} &= e^{-\mu it\sqrt{n}/\sigma}e^{\mu it\sqrt{n}/\sigma}e^{-\mu_2' t^2/2\sigma^2 - (it\mu_1')^2/2\sigma^2} \\ &= e^{-t^2/2},\end{aligned}$$

so $\bar{x}'$ is asymptotically distributed as $N(0,1)$, and therefore $\bar{x}$ is asymptotically distributed as $N(\mu, \sigma^2/n)$. This is a special case of the famous *central limit theorem.*[1] It does not hold for distributions that lack a variance, such as the Cauchy distribution.

> **Problem 29.** Show that the distribution of the mean of a sample drawn from a Cauchy distribution is identical to the distribution of a single observation, and hence that its sampling variance is infinite. (By symmetry it is unbiased for the centre of symmetry.)

The central limit theorem is more general than this statement that the sample mean of a distribution that possesses moments follows a Gaussian distribution. A more complete statement would be that the sum of n variables (not necessarily all drawn from the same distribution) follows a Gaussian distribution as $n \to \infty$, providing that all possess a second moment and that the ratio of second moments of any two variables is bounded. This last condition means that all of the variables contribute to the sum; if a few terms dominate the sum, then the effective number of variables doesn't become infinite and it is not surprising that the theorem doesn't hold.

You should not think that all distributions become Gaussian for large n. The central limit theorem applies to the distributions of sums of variables, rather than being a vague and general result about all of statistics. It is, however, no coincidence that the large n limit of a Poisson distribution is Gaussian. You showed in problem 6 that the sum of two independent Poisson variables is Poisson; this allows us to think of a large n Poisson distribution as being the sum of a number of independent variables, which in turn entitles us to apply the central limit theorem.

Problem 30. A prosperous farmer, inspired by a book on statistics, weighs all of his sheep and goats. Can he reasonably expect the goats' weights to be normally distributed? Can he expect the distribution of the weights of all of his animals to be Gaussian?

6.1. Distribution of s^2 if $x = N(0, \sigma^2)$

For a Gaussian (but *only* for a Gaussian) it is possible to show that $\bar{x}$ and s^2 are independent.[2]

Problem 31. For the case $n = 2$ calculate the joint p.d.f of $\bar{x}$ and s^2, and show that it is the product of the p.d.f.'s for $\bar{x}$ and s^2. Indicate how you would generalize this derivation for a sample of size n.[3] (*Hint:* write down the cumulative p.d.f. of $\bar{x}$ and s^2 as a double integral over x_1 and x_2 [both $N(0, 1)$ variables]. Change variables and differentiate to obtain the p.d.f.)

Problem 32. Consider a sample of n points drawn from a population with p.d.f. f and characteristic function ϕ. Let

$\psi(t_1, t_2)$ be the joint characteristic function of $\bar{x}$ and s^2, i.e.,

$$\psi(t_1, t_2) = \left\langle e^{it_1\bar{x}} e^{it_2 s^2} \right\rangle,$$

and show that

$$\left.\frac{\partial \psi(t_1, t_2)}{\partial t_2}\right|_{t_2=0} =$$

$$i\left(\frac{n-1}{n}\right)\phi^{n-2}(t_1/n)\left(\phi'^2(t_1/n) - \phi(t_1/n)\phi''(t_1/n)\right),$$

where $\phi' \equiv d\phi/dt$. If $\bar{x}$ and s^2 are independent, show that

$$\left.\frac{\partial \psi(t_1, t_2)}{\partial t_2}\right|_{t_2=0} = i\phi^n(t_1/n)\frac{n-1}{n}\sigma^2$$

and hence show that f must be Gaussian.[4]

Using this result we discover that we did almost all of the work required to find the distribution of s^2 when we calculated the distribution of the sum of the squares of n $N(0,1)$ variates in section 4.1. Consider

$$\frac{\sum x^2}{\sigma^2} = \frac{n}{\sigma^2}\left(s^2 + \bar{x}^2\right).$$

The left-hand side is the sum of the squares of n independent $N(0,1)$ variates, and thus must follow a χ_n^2 distribution. From the previous section we know that $\bar{x}$ is an $N(0, \sigma^2/n)$ variable, so $n\bar{x}^2/\sigma^2$ is the square of an $N(0,1)$ variable, i.e., it follows a χ_1^2 distribution. Because s and $\bar{x}$ are independent we see that

$$\frac{ns^2}{\sigma^2}$$

follows a χ_{n-1}^2 distribution, and therefore that

$$dF = \frac{n^{(n-1)/2}}{(2\sigma^2)^{(n-1)/2}(n/2 - 3/2)!} e^{-ns^2/2\sigma^2} s^{n-3}\, ds^2.$$

Problem 33. Show that we can write $\sum (x - \bar{x})^2$ in the form $\mathbf{x}^T A \mathbf{x}$, where $A^2 = A$ and $\mathrm{Tr}(A) = n - 1$. Deduce that ns^2/σ^2 is a χ_{n-1}^2 variable.

Problem 34. Show that s^2's mean and variance are indeed $(n-1)\sigma^2/n$ and $2(n-1)\sigma^4/n^2$.

Problem 35. What are the mean and variance of s? What are the large n limits of these results? (*Hint:* use Stirling's formula to expand the factorials, and derive a version of the lemma of section 3.5 correct to $O(1/n)$.)

6.2. Student's t-Distribution

As we shall see when we come to discuss hypothesis testing, we are often interested in the distribution of the ratio of the mean to the standard deviation of a sample drawn from a Gaussian population $N(\mu, \sigma^2)$. We know that $\bar{x}$ follows an $N(\mu, \sigma^2/n)$ distribution, so you might think that $(\bar{x} - \mu)/(s/\sqrt{n-1})$ would follow an $N(0, 1)$ distribution, but this is not true. Because, as you showed in problem 31, s^2 and $\bar{x}$ are independent for a sample drawn from a Gaussian population, it is straightforward to derive the actual distribution of $(\bar{x} - \mu)/s$; in general it is not so simple.

The distributions of $\bar{x}$ and s are[5]

$$dF_{\bar{x}} \propto e^{-n(\bar{x}-\mu)^2/2\sigma^2}\, d\bar{x}$$

and

$$dF_s \propto e^{-ns^2/2\sigma^2}\, s^{n-2}\, ds$$

(neglecting mere coefficients).

Consider $z = (\bar{x} - \mu)/s$, then[6]

$$dF_z = \int_0^\infty f_s(s)\, ds \int_{zs}^\infty f_{\bar{x}}(\bar{x})\, d\bar{x}\, dz,$$

so $f_z = dF_z/dz = dF_z/d(zs) \cdot d(zs)/dz$, i.e.,

$$\begin{aligned} f_z &= \int_0^\infty f_s(s) f_{\bar{x}}(zs) s\, ds \\ &\propto \int_0^\infty e^{-ns^2/2\sigma^2} s^{n-2} e^{-nz^2s^2/2\sigma^2} s\, ds \\ &\propto \int_0^\infty e^{-(1+z^2)ns^2/2\sigma^2} s^{2(n/2-1)}\, ds^2, \end{aligned}$$

or, changing variables with $y = (1 + z^2)ns^2/2\sigma^2$,

$$f_z = \left(\frac{2\sigma^2}{(1+z^2)}\right)^{n/2} \int_0^\infty e^{-y} y^{n/2-1}\, dy.$$

The integral can be done (it's $(n/2 - 1)!$), but as it is independent of z in any case it can be treated like any other constant

$$dF_z = C(1 + z^2)^{-n/2}\, dz;$$

or, requiring that the integral of f_z over $[-\infty, \infty]$ be 1,

$$dF_z = \frac{(n/2-1)!}{\pi^{1/2}(n/2-3/2)!} \frac{dz}{(1+z^2)^{n/2}}.$$

Let us change variables to t, defined as

$$\begin{aligned} t &= \frac{\bar{x}-\mu}{\left(\frac{1}{n-1}\sum(x-\bar{x})^2\right)^{1/2}/\sqrt{n}} \\ &= (n-1)^{1/2} z, \end{aligned}$$

so, writing $\nu \equiv n-1$, we have

$$dF_t = \frac{(\nu/2-1/2)!}{(\nu\pi)^{1/2}(\nu/2-1)!} \frac{1}{(1+t^2/\nu)^{(\nu+1)/2}} dt,$$

which is known as *Student's t-distribution* with ν degrees of freedom. See section 4.3 for a discussion of its properties; unsurprisingly, one of these is that its large n limit is a Gaussian.

More formally we can say that if y is an $N(0, \sigma^2)$ variate and s^2 is an unbiased estimate of σ^2, if y and s^2 are independent, and if s^2 is a multiple of a χ^2_ν variable, then y/s follows a t_ν-distribution. In the case that we have been considering we know that $y = \bar{x} - \mu$ follows a $N(0, \sigma^2/n)$ distribution, that $ns^2/(n-1)$ is an unbiased estimator for σ, that $\langle \bar{x}s^2 \rangle = 0$, and that $s^2 = \sigma^2\chi^2_{n-1}$, so all of these conditions are satisfied.

We shall see an alternative derivation of the t-distribution from a Bayesian viewpoint in section 8.3.

As we shall soon see, we are often interested in knowing how many standard deviations a given value is from the mean; if the parent population is Gaussian (and we have estimated both mean and variance from the sample) this is a question about the t-distribution.

6.3. Sampling from a Finite Population

We have been implicitly assuming that our population was infinite and defined by a probability density function. How are our conclusions changed if the parent is finite, with N members?

If we have a sample of size n the mean is of course still[7]

$$\bar{x} = \frac{1}{n}\sum_{i=1}^{n} x_i$$

and we can calculate its expectation value:

$$\begin{aligned}\langle \bar{x} \rangle &= \left\langle \frac{1}{n} \sum x_i \right\rangle \\ &= \frac{1}{n} \sum \langle x_i \rangle \\ &= \langle x_i \rangle .\end{aligned}$$

We must evaluate the expectation value over our finite population:

$$\begin{aligned}&= \frac{1}{N} \sum_{i=1}^{N} x_i \\ &= \mu,\end{aligned}$$

so $\langle \bar{x} \rangle = \mu$ as usual. What about the sample variance?[8]

$$\begin{aligned}s^2 &= \frac{1}{n} \sum_{i=1}^{n} x_i^2 - \left(\frac{1}{n} \sum_{i=1}^{n} x_i \right) \\ &= \frac{1}{n} \sum_{i=1}^{n} x_i^2 - \frac{1}{n^2} \left(\sum_{i=1}^{n} x_i^2 + \sum_{i \neq j}^{n} x_i x_j \right),\end{aligned}$$

so its expectation value is

$$\left\langle s^2 \right\rangle = \frac{n-1}{n} \left(\left\langle x_i^2 \right\rangle - \left\langle x_i x_j \right\rangle_{i \neq j} \right).$$

This result holds whatever the size of the population. We can evaluate the expectation values for our finite population:

$$\begin{aligned}\left\langle x_i^2 \right\rangle &= \frac{1}{N} \sum_{i=1}^{N} x_i^2 \\ &= \sigma^2 + \mu^2\end{aligned}$$

and

$$\begin{aligned}\left\langle x_i x_j \right\rangle_{i \neq j} &= \frac{1}{N(N-1)} \sum_{i \neq j}^{N} x_i x_j \\ &= \frac{1}{N(N-1)} \left(\left(\sum_{i=1}^{N} x_i \right)^2 - \sum_{i=1}^{N} x_i^2 \right) \\ &= \frac{1}{N(N-1)} \left(N^2 \mu^2 - N(\sigma^2 + \mu^2) \right) \\ &= \mu^2 - \frac{\sigma^2}{N-1},\end{aligned}$$

so

$$\langle s^2 \rangle = \frac{N}{N-1}\frac{n-1}{n}\sigma^2.$$

When N is large this reduces to the usual formula; when $n = N$ we have sampled every member of the population and it is hardly surprising that $s^2 = \sigma^2$.

Problem 36. What is $\bar{x}$'s variance?

6.4. Asymptotic Distribution of the Median

Rather surprisingly, it is possible to derive an expression for the probability distribution of the median, in the limit of large samples.[9] (The argument is easily generalized to deal with any quantile of the distribution.)

Consider a sample of size $n = 2\nu + 1$ taken from a population with p.d.f. dF/dx. The probability that ν members of the sample lie below x_m; that there is 1 member in the range $x_m, x_m + dx$; and that the remaining ν lie above x_m (i.e., the probability that the median of the sample is x_m) is

$$g(x_m)\,dx_m = F(x_m)^\nu \cdot f(x_m)\,dx_m \cdot (1 - F(x_m))^\nu,$$

and g is the p.d.f. of the median.

Where is the maximum of g? Taking logs and differentiating, we find that the maximum occurs where

$$\frac{\nu}{F}f + \frac{f'}{f} - \frac{\nu}{1-F}f = 0$$

(remember that $f = dF/dx$). For large n the f'/f term is much smaller than the terms in ν, and we see that the maximum is at $F(x_m) = 1/2$, a not unexpected result.

Let us now consider the distribution of $\zeta = x_m - \xi_m$, where ξ_m is the true median.

$$\begin{aligned} F(\xi_m + \zeta) &= F(\xi_m) + \left.\frac{dF}{dx_m}\right|_{x_m=\xi_m}\zeta + \frac{1}{2}\left.\frac{d^2F}{dx_m^2}\right|_{x_m=\xi_m}\zeta^2 \\ &= \frac{1}{2} + f(\xi_m)\zeta + \frac{1}{2}f'(\xi_m)\zeta^2, \end{aligned}$$

so we can write g as

$$g(x_m) = (1/2 + f(\xi_m)\zeta + \frac{1}{2}f'(\xi_m)\zeta^2)^\nu f(x_m)(1/2 - f(\xi_m)\zeta - \frac{1}{2}f'(\xi_m)\zeta^2)^\nu$$
$$= f(x_m)((1/2)^2 - (f(\xi_m)\zeta)^2)^\nu$$
$$= C(1 - (2f(\xi_m)\zeta)^2)^\nu,$$

and, as $n \to \infty$, this becomes

$$g(x_m) \sim e^{-\nu(2f(\xi_m)\zeta)^2}$$
$$= e^{-n(2f(\xi_m)\zeta)^2/2}.$$

We see that the distribution of the median is asymptotically $N(\xi_m, 1/(4f^2(\xi_m)n))$. For a Gaussian $N(\mu, \sigma^2)$ the median $\xi_m = \mu$, so $f(\xi_m) = 1/(\sqrt{2\pi}\sigma)$ and the variance of the median is $\pi\sigma^2/2n = (1.253)^2\sigma^2/n$.

Problem 37. Show that the variance of the pth percentile of a distribution is asymptotically $p(1-p)/nf_1^2$, where f_1 is the value of the p.d.f. at the pth percentile (note that this reduces to the previous result for the median when $p = 1/2$).

Problem 38. For a Cauchy distribution, show that the large n limit of the median's sampling variance is $\pi^2/4n$, and therefore that the median is a consistent estimator of the distribution's central point (by symmetry, it is unbiased).

6.5. Bootstraps and Jackknives

Let us consider a slightly different way of estimating a population's properties.[10] We have a sample of n independent points $\{x_i\}$ from the same distribution, so we can estimate their distribution function as[11]

$$f^* = \frac{1}{n}\sum_i \delta(x - x_i)$$

(this choice of f^* maximizes $\Pr(x_i|f^*)$, so if you've skipped ahead to the discussion of maximum likelihood estimators you'll know that this is the ML estimator of f). We can use this estimated p.d.f. to calculate population properties, for example the mean:

$$\langle x \rangle^* = \int_{-\infty}^{\infty} x f^* \, dx = \frac{1}{n} \sum_i x_i,$$

where $\langle \cdots \rangle^*$ indicates an expectation taken over f^* rather than the (unknown) f; such a use of f^* leads to what are known as *bootstrap estimates.* We can investigate f^*'s properties in detail; for example, we can imagine drawing many samples from f^* and looking at the distribution of the sample's means. We would, of course, rather draw the samples from f itself, but regrettably that is not possible.

If f is a discrete distribution with only a finite number of possible values, it is clear that $f^* \to f$ as $n \to \infty$, so what we learn about f^* may reasonably be taken to apply to f. If f is continuous this is less obvious, but it may be shown that statistics have the same asymptotic distribution under f^* as under f;[12] the convergence is about as fast as that involved in invoking the central limit theorem.

This use of f^* to estimate properties of the population is not very different from the familiar procedure of assuming a functional form for f (e.g., $N(\mu, \sigma^2)$), estimating the parameters from the $\{x_i\}$, and then calculating any properties of the population that interest us. In fact, this procedure is sometimes referred to as a *parametric bootstrap.*

Any sample we draw from f^* will have values drawn from the original $\{x_i\}$, so all that we need specify is how many copies n_i we have of each. The n_i follow a multinomial distribution with $p_i = 1/n$, so problem 16 tells us that

$$\langle n_i \rangle^* = 1$$

and

$$\langle n_i n_j \rangle^* = 1 + \delta_{ij} - 1/n.$$

The mean of a sample is $\bar{x} = 1/n \sum n_i x_i$, so $\langle \bar{x} \rangle^* = 1/n \sum x_i$ and we see that $\bar{x}$ is unbiased (with respect to f^*). What about its variance?

$$\begin{aligned} V^*(\bar{x}) &= \langle \bar{x}^2 \rangle^* - \langle \bar{x} \rangle^{*2} \\ &= \frac{1}{n^2} \sum_{ij} \langle n_i n_j \rangle^* x_i x_j - \bar{x}^2 \end{aligned}$$

$$= \frac{1}{n^2}\sum_i x_i^2 + \left(1 - \frac{1}{n}\right)\sum_{ij}\left(\frac{1}{n}\sum_i x_i\right)^2 - \bar{x}^2$$

$$= \frac{1}{n}\left(\frac{1}{n}\sum_i x_i^2 - \bar{x}^2\right)$$

$$= \frac{1}{n}s^2,$$

which is nearly the usual result (but not quite, as s^2 isn't unbiased for σ^2).

If I were interested in some statistic t that was much more complex than $\bar{x}$ (a simple example would be $(3 \times \text{median} - \text{mode})/2$; cf. problem 2) such a calculation would be difficult or impossible, but we can use a Monte Carlo simulation to estimate $\langle t \rangle^*$: Draw a sample from f^* — i.e., draw n points from $\{x_i\}$ *with replacement* — and calculate t; repeat this procedure B times and use the resulting set of values to calculate biases, variances, and other statistical desiderata.

There seems to be a certain amount of confusion about the approximations involved in bootstrap calculations. The first approximation, that $f^* \sim f$, is unavoidable and inherent in the method, but the second, the Monte Carlo calculation of $\langle f \rangle^*$, is solely a matter of convenience. When the analytical analysis is tractable no approximation is needed, and in any case if we had sufficient computer time available we could choose each of the $n!$ possible samples from $\{x_i\}$ and evaluate f^*'s properties exactly. This would be foolish, of course, as all that we need do is ensure that the errors arising from the second approximation are much smaller than those due to the first.

Problem 39. In section 12.3 we shall find an approximate expression for the p.d.f. of Pearson's correlation coefficient r_{xy}, but here let us attack the same problem with a bootstrap. Write a program to draw a sample of n points $\{x_i\}$ from a bi-Gaussian distribution with specified correlation coefficient ρ_{xy}, and find the sample correlation coefficient r_{xy}. Then draw B bootstrap samples from $\{x_i\}$, calculate r^*_{xy} from each, and hence estimate the bias and variance of r_{xy} as an estimate of ρ_{xy}. If $\rho_{xy} = 0.5$, $n = 20$, and $B = 100$, how do these compare with the theoretical estimates $r - \rho \sim -\rho(1 - \rho^2)/2n$ and $V(r) \sim (1 - \rho^2)^2/n$? (Note that you can generate a bi-Gaussian sample by drawing two independent Gaussian samples and rotating axes.)

An apparently different way of estimating sample bias and variance is the *Jackknife*,[13] so named from its claimed robustness and general applicability. We shall soon see that the jackknife is closely related to the bootstrap.

Consider a consistent statistic t and write the value of t calculated from a sample of size n as t_n,[14] so t_∞ is the population value. Expand t_n as a series in n^{-1} (note the resemblance to Richardson extrapolation; for example, Romberg integration[15]):

$$t_n = t_\infty + n^{-1}t'_n + O(n^{-2})$$

i.e.,

$$\begin{aligned} t_\infty &= nt_n - (n-1)t_{n-1} + O(n^{-2}) \\ &= t_n + (n-1)(t_n - t_{n-1}) + O(n^{-2}). \end{aligned}$$

We have a single sample of n points, from which we can extract n subsamples of size $n-1$ by omitting one element; let $t_{n-1,i}$ have the ith element omitted. If I find $t_{n-1,i}$ for each subsample and average to give $\bar{t}_{n-1}$, I can approximate t_{n-1} by $\bar{t}_{n-1}$ and define a bias-corrected jackknife estimate as

$$t_n^J \equiv t_n + (n-1)(t_n - \bar{t}_{n-1}).$$

It t_n has variance of order $(1/n)$, t_n^J has the same asymptotic variance as t_n; if you're tempted to calculate higher-order jackknives (based on omitting two or more elements), be warned that this isn't generally true.

We can also use $t_{n-1,i}$ to estimate t_n's variance as

$$V^J(t_n) \equiv \frac{n-1}{n} \sum_i (t_{n-1,i} - \bar{t}_{n-1})^2.$$

Problem 40. If $t_n = \bar{x}$, show that $t_n^J = \bar{x}$.

Problem 41. If $t_n = (1/n)\sum(x-\bar{x})^2$, show that $t_n^J = (1/(n-1))\sum(x-\bar{x})^2$; note that this is indeed unbiased.

Problem 42. Show that $V^J(t_n) = s^2/n$ if $t_n = \bar{x}$ (where s^2 is now an unbiased estimate of σ^2).

Problem 43. Modify your program from problem 39 to calculate jackknife estimates of bias and variance.

It may be shown that[16]

$$\left\langle \sum_i (t_{n-1,i} - \bar{t}_{n-1})^2 \right\rangle \geq V_{n-1},$$

where V_{n-1} is the true variance for a sample of size $n - 1$; the factor $(n - 1)/n$ in the definition of V^J is introduced to scale the variance to a sample of size n, as many common statistics obey the relation $V_n = (n - 1)/nV_{n-1} + O(1/n^3)$; the mean is a familiar example.

Suspicious readers will not have been convinced by my derivation of the jackknife correction for bias,[17] as I could just as well have written

$$t_n = t_\infty + (n + 1)^{-1}t'_n + O(n^{-2}),$$

leading to

$$t_\infty = t_n + n\,(t_n - \bar{t}_{n-1}).$$

The real justification for our definition of t_n^J is the calculation in problem 41, wherein the variance's bias is indeed removed.

There is a close connection between jackknives and bootstraps (despite the mixed metaphor).[18] As we saw, to choose a bootstrap sample we specify a set of integers $\{n_i\}$ satisfying $\sum n_i = n$, so we may think of a statistic t as being a function defined in an n-dimensional space, and $\{n_i\}$ as a vector $\mathbf{n}$ in this space. If we relax the requirement that the n_i be integral we can bring the jackknife into the same framework; the value of $t_{n-1,j}$ is t evaluated at the point $\mathbf{n}^{J,j}$ given by

$$n_i = \begin{cases} n/(n-1), & \text{if } i \neq j, \\ 0, & \text{if } i = j. \end{cases}$$

Let us approximate the statistic t by the linear approximation t^L, which equals t at the points $\mathbf{n}^{J,i}$, and proceed to find t^L's properties under bootstrap sampling:

$$\begin{aligned} t^L &= \frac{n-1}{n} \sum_i t_{n-1,i} \left(\frac{n}{n-1} - n_i \right) \\ &= n\bar{t}_{n-1} - \frac{n-1}{n} \sum_i n_i t_{n-1,i} \\ &= \bar{t}_{n-1} + \frac{n-1}{n} \sum_i (1 - n_i) t_{n-1,i}, \end{aligned}$$

so

$$\begin{aligned}\langle t^L\rangle^* &= \bar{t}_{n-1} + \frac{n-1}{n}\sum_i(1-\langle n_i\rangle^*)t_{n-1,i}\\ &= \bar{t}_{n-1}.\end{aligned}$$

Now the variance:[19]

$$\begin{aligned}V^*(t^L) &= \left(\frac{n-1}{n}\right)^2\sum_{ij}\langle(1-n_i)(1-n_j)\rangle^* t_{n-1,i}t_{n-1,j}\\ &= \left(\frac{n-1}{n}\right)^2\sum_{ij}(1-\delta_{ij})t_{n-1,i}t_{n-1,j}\\ &= \left(\frac{n-1}{n}\right)^2\sum_{i}(t_{n-1,i}-\bar{t}_{n-1})^2\\ &= \frac{n-1}{n}V^J(t),\end{aligned}$$

so the jackknife variance is a linear approximation to the bootstrap variance, modified by a factor that makes it unbiased in some simple cases.

To find the jackknife bias correction we have to work to higher order.[20] The next problem shows that if we define t^Q as a quadratic surface that passes through the observed data point ($t(\mathbf{1})$, i.e., where $\mathbf{n} = \mathbf{1}$) and the n points $t(\mathbf{n}^{J,i})$, then the bootstrap bias correction $\text{Bias}^* \equiv \langle t^Q\rangle^* - t(\mathbf{1})$ is equal to $(n-1)/n\,\text{Bias}^J$.

Problem 44. Show that $\text{Bias}^* = (n-1)/n\,\text{Bias}^J$.

It is not surprising to learn that the jackknife variance proves more reliable than the jackknife bias correction, as it is based on a simpler approximation; the inequality quoted without proof above gives further confidence in its usefulness. If a statistic's jackknife bias is a good deal smaller than the statistics's standard deviation (maybe estimated using a jackknife or bootstrap), it is probably reasonable to assume that bias is not a major problem.

You can doubtless think of other jackknife approximations.[21] The most obvious is probably to approximate t not as t^L but as the hyperplane tangent to t at the observed data point; this is known as the *infinitesimal jackknife* and doesn't seem to be especially useful.

References

1: *K&S* 7.26
2: *K&S* Exc. 11.19
3: *K&S* 11.2, 11.3
4: *K&S* Exc. 11.19
5: *K&S* Ex. 11.7
6: *K&S* Ex. 11.8, 16.10
7: *K&S* 12.20
8: *K&S* 12.20
9: *K&S* 10.10, 11.4
10: $K\&S_V$ 17.10
11: *Efron* 5.0
12: *Efron* 5.6
13: *Efron* 2.0
14: *K&S* 17.10
15: *NR* 4.3
16: *Efron* 4.0
17: *Efron* 2.6
18: *Efron* 6.1
19: *Efron* 6.2
20: *Efron* 6.6
21: *Efron* 6.3

7. Bayes' Theorem and Maximum Likelihood

Bayes' theorem and the likelihood play an important part in statistics. This section merely serves to introduce them; we'll return to them in a more formal way in Chapter 10.

7.1. Bayes' Theorem and Postulate

It is obvious that, for two events A and B, the probability of both happening (written as $P(A \cap B)$) is given by[1,2]

$$P(A \cap B) = P(A|B)P(B) = P(B|A)P(A),$$

so

$$P(B|A) = \frac{P(A|B)P(B)}{P(A)}$$

(where $P(A|B)$ is the probability of A if we already know that B has occurred and is pronounced "the probability of A given B"). This is known as *Bayes' theorem*, and is not controversial. For example, A could be some observation and B some set of model parameters. $P(B)$ is known as a *prior probability*, $P(B|A)$ is a *posterior probability*, and $P(A|B)$ is called the *likelihood.*[3]

Bayes' postulate, as opposed to his theorem, consists of the claim (in the absence of other knowledge) that all prior probabilities, or "priors," should be treated as equal.[4] Although this might seem innocuous at first sight, a little reflection reveals that even if the probability of θ is uniform, the probability of $f(\theta)$ is not. Reasonably compelling arguments can be given for choosing a constant prior for location statistics such as the mean[5] (i.e., adopting Bayes' principle), and for choosing a prior proportional to the reciprocal for scale statistics such as the variance[6] (i.e., $p(\sigma^2) \propto 1/\sigma^2$); see the next problem for one such set of arguments. For large samples the posterior probability is dominated by the likelihood anyway, so the choice of prior is less important.

7.2. Maximum Likelihood Estimators

We can use Bayes' theorem to estimate the parameters of a model from a sample. Interpreting B as the model parameters and A as the data, and accepting Bayes' postulate that $P(B)$ is a constant, all I have to do to maximize the probability that my estimate of B is correct is to maximize the total likelihood of the model (although $P(A)$ is unknown it is also a constant). The resulting value of B is known as a *maximum likelihood estimate* (MLE).[7] We could have chosen any statistic of the posterior probability such as its median or mean instead of its maximum (i.e., the mode); we chose the latter because it's relatively easy to find and because it's easy to prove theorems about its properties.

As a concrete example, consider the problem of estimating the mean of a sample of size n from a Gaussian population $N(\mu, \sigma^2)$, where σ is known.[8] We'll accept Bayes' postulate and assume that all values of μ are equally likely. The likelihood L is the probability that we'd get the particular sample that we have got, if we knew that the population mean was some particular value of μ; it's given by

$$L = \frac{1}{(2\pi)^{n/2}\sigma^n} \prod_i e^{-(x_i-\mu)^2/2\sigma^2}.$$

It is usually easier to work with $\ln L$:

$$\ln L = -\frac{n}{2}\ln(2\pi) - n\ln\sigma - \sum_i \frac{(x_i - \mu)^2}{2\sigma^2},$$

and differentiating to find the maximum gives

$$\frac{\partial \ln L}{\partial \mu} = 0 = \frac{1}{\sigma^2}\sum_i (x_i - \mu),$$

so the MLE for μ is

$$\hat{\mu} = \frac{1}{n}\sum_i x_i,$$

a not unexpected result (we write $\hat{\theta}$ as the estimator of θ).

How about estimating σ? Differentiating $\ln L$ with respect to σ gives

$$\frac{\partial \ln L}{\partial \sigma} = 0 = -\frac{n}{\sigma} + \frac{1}{\sigma^3}\sum_i (x_i - \mu)^2,$$

so

$$\hat{\sigma}^2 = \frac{1}{n}\sum_i (x - \mu)^2;$$

again, not a totally unexpected result. In fact, σ^2 cannot take on all possible values as it cannot be negative, so it is sometimes argued that the proper probability distribution for σ is $d\sigma/\sigma$ (i.e., $P(B) = 1/\sigma$), in which case the MLE estimator becomes

$$\hat{\sigma}^2 = \frac{1}{n+1}\sum_i (x-\mu)^2.$$

As $n \to \infty$ the distinction becomes less and less important; we can think of the greater and greater amount of information in the sample drowning out the information in the prior.

Problem 45. Let the prior probability of σ be $p(\sigma)$. Write down the posterior probability of σ if the x_i are drawn from an $N(0, \sigma^2)$ population. By arguing that this probability shouldn't be affected by a transformation of the form $\mathbf{x} \to \alpha\mathbf{x}$, $\sigma \to \alpha\sigma$ argue, that $p(\sigma) = 1/\sigma$.

Problem 46. What is the mean value of σ^2 (i.e., evaluated from the posterior probability of the previous problem)?

Problem 47. If I make one measurement n of a Poisson variable x, what is the ML estimate of x's mean μ?

Problem 48. If I have to wait 10 minutes for my bus, what is the maximum likelihood estimate of the time between buses?

References

1: *K&S* 8.2

2: *Lee* 1.1

3: *K&S* 8.3

4: *K&S* 8.4

5: *Lee* 2.5

6: *K&S* 8.14, 21.28

7: *K&S* 8.6, 18.1

8: *K&S* Ex. 8.5

8. Confidence Intervals

8.1. Introduction

Sometimes we know (or can estimate) the properties of a population, and want to know if a certain observation is drawn from it. For example, we might have measured the velocity dispersion of a certain cluster of galaxies, and now want to know if an additional galaxy is a cluster member. Let us call the population's p.d.f. $f(x)$, and the extra observation x_0.

It doesn't make much sense (at least for a continuous distribution) to ask for the probability of obtaining a value of exactly x_0, as it is given by $f(x_0)\,dx$ and vanishes as $dx \to 0$. But we *can* ask for the probability that we'd measure a value of x at least as large as x_0:[1]

$$P(x > x_0) = \int_{x_0}^{\infty} f(x)\,dx$$
$$= 1 - \int_{-\infty}^{x_0} f(x)\,dx.$$

If the mean of f is μ, I might alternatively want the probability of obtaining a value of x at least as far from the mean as x_0:

$$P(x \text{ is further from mean than } x_0) = 1 - \int_{2\mu - x_0}^{x_0} f(x)\,dx.$$

We can turn this argument around and specify not x_0 but a desired probability. For example, if $f(x)$ is a Gaussian $N(\mu, \sigma^2)$, I can say that the probability of making an observation larger than $\mu + 1.96\sigma$ or smaller than $\mu - 1.96\sigma$ is 5% because

$$\frac{1}{\sqrt{2\pi}\sigma} \int_{\mu - 1.96\sigma}^{\mu + 1.96\sigma} e^{-(x-\mu)^2/2\sigma^2}\,dx = 0.95.$$

If I claim that any observation made from a distribution will lie in the range $[\xi_1, \xi_2]$ I will be wrong in a fraction α of cases if [2]

$$\int_{\xi_1}^{\xi_2} f(x)\,dx = 1 - \alpha.$$

This range is called a *confidence interval*, at a confidence level $1 - \alpha$. For a Gaussian, $[\mu - 1.96\sigma, \mu + 1.96\sigma]$ is a 95% confidence interval for an observation.

This procedure may not seem very useful, but in fact it is one of the commonest procedures in statistics. For example, suppose that the parent population is known to be $N(\mu, \sigma^2)$, where σ is known but μ is not. We have a sample of size n, and wish to estimate μ. How should we proceed? We can calculate the sample mean $\bar{x}$, and we know that it is unbiased for μ, but how good an estimate is it? The distribution of $\bar{x}$ is $N(\mu, \sigma^2/n)$, so we can say (at a 95% confidence level) that the population mean lies in the range $[\bar{x} - 1.96\sigma/\sqrt{n}, \bar{x} + 1.96\sigma/\sqrt{n}]$, because if this weren't true I'd have a probability of less than 5% of drawing my sample from the parent distribution.

> **Problem 49.** What is the 99% confidence interval for the population mean, given a sample of size n drawn from a Gaussian population with standard deviation σ?

In reality we seldom know the variance of the parent population, although we may be happy to assume that it is Gaussian. In this case we have to estimate both μ and σ from the sample, and additional difficulties arise.

When we knew σ we were able to express our confidence interval in terms of $(\bar{x} - \mu)/(\sigma/\sqrt{n})$, which we know to be an $N(0, 1)$ variate. We might be tempted to use the same procedure, and base our interval upon $(\bar{x} - \mu)/(s/\sqrt{n-1})$.

> **Problem 50.** Why $\sqrt{n-1}$, not $\sqrt{n}$?

Because we are estimating both $\bar{x}$ and s, $(\bar{x} - \mu)/(s/\sqrt{n-1})$ is not an $N(0, 1)$ variate, except in the limit $n \to \infty$.[3] Fortunately we know what its distribution is, namely a t-distribution with $n - 1$ degrees of freedom. Student's t is well tabulated (and easy to calculate),[4] so it is easy to specify confidence intervals.

> **Problem 51.** What is the 95% confidence interval for the population mean of a sample of size 10 drawn from a Gaussian population of unknown mean and variance?

A statistic such as t is said to be *pivotal* if its distribution doesn't depend on any parameters of the model; for example, if we had tried to base a test on $\bar{x} - \mu$ instead of t, the unknown σ would have entered into its sampling distribution.

Problem 52. Find a statistic that is pivotal for the mean μ of a Poisson distribution, given a sample of size n.

8.2. The Choice of Confidence Intervals

We have been choosing confidence intervals that are symmetrical about the mean, but this is not the only possible choice. If we are interested if a variable is too small, but we don't care if it's too large (the classical example is quality control in a factory; you won't be sued if you give your customers 17oz. of biscuits in a 1lb. box, but you'd better not give them only 15oz.), we might be more interested in a *1-sided* interval[5] and claim that x lies in the interval $[-\infty, \xi]$, where

$$\int_{\xi}^{\infty} f(x)\, dx = \alpha.$$

Problem 53. What is the 1-sided 95% confidence interval for the population mean, given a sample of size n drawn from a Gaussian population with standard deviation σ?

Even for a *2-sided* interval such as those that we have been discussing, it is not always obvious how to choose the values of ξ_1 and ξ_2.[6] For the case of a symmetrical distribution such as the Gaussian it's reasonably obvious that we should take $|\mu - \xi_1| = |\mu - \xi_2|$, but for a skew distribution we might want to replace this by the condition

$$\int_{-\infty}^{\xi_1} f\, dx = \int_{\xi_2}^{\infty} f\, dx = \alpha/2,$$

which will not, in general, be equivalent. Another alternative would be to require that the allowed range of x values be made as small as possible, i.e., to choose the ξ's to minimize $\xi_2 - \xi_1$, and in general this won't be equivalent either (an example of a case where it isn't is a χ^2 distribution); yet another approach would be to choose a region that satisfies $f(x) > c_\alpha$, with c_α a constant determined by the significance level.[7]

The problem of choosing intervals is not always easy; in fact, it isn't always even possible.[8] Consider an $N(\mu, \sigma^2)$ population where σ is known but μ isn't, and from which we have been given a sample of size n. The statistic $X^2 = \sum(x - \mu)^2/\sigma^2$ follows a χ^2 distribution with n degrees of freedom, and I can find a pair of values X_0^2 and X_1^2 such that $P(X_0^2 \le X^2 \le$

$X_1^2) = 1 - \alpha$ for any given α. Of course I don't know μ, but I can write down the identity

$$\begin{aligned}\sigma^2 X^2 &= \sum (x-\mu)^2 \\ &= \sum ((x-\bar{x}) + (\bar{x}-\mu))^2 \\ &= \sum (x-\bar{x})^2 + 2(\bar{x}-\mu)\sum (x-\bar{x}) + n(\bar{x}-\mu)^2 \\ &= n(s^2 + (\bar{x}-\mu)^2).\end{aligned}$$

I can now rewrite the condition on the allowed range of X^2 as

$$P(\sigma^2 X_0^2/n \le s^2 + (\bar{x}-\mu)^2 \le \sigma^2 X_1^2/n) = 1 - \alpha,$$

i.e.,

$$P(\sigma^2 X_0^2/n - s^2 \le (\bar{x}-\mu)^2 \le \sigma^2 X_1^2/n - s^2) = 1 - \alpha.$$

It is entirely possible that one day a sample will be found that has $s^2 > \sigma^2 X_1^2/n$, and this inequality will duly inform us that $(\bar{x}-\mu)^2$ lies between two negative numbers, which appears implausible. We are not of course required to imagine a complex population mean; we have merely found a sample that would only occur with a probability of less than α, but we have not succeeded in placing a limit on μ. We have devised a test that depends on the distribution of both $\bar{x}$ and s^2 and tried to assign all of the uncertainty to just one of them, $\bar{x}$. Our problems stem from our choice of an inappropriate statistic X^2,†and if we'd chosen better we wouldn't have had any problems. For example, we know that $\bar{x}$ is distributed as $N(\mu, \sigma^2/n)$, so we could have simply used the usual procedure to place constraints on $\bar{x} - \mu$ and deduced that

$$P(-\beta\sigma/\sqrt{n} \le \bar{x} - \mu \le \beta\sigma/\sqrt{n}) = 1 - \alpha$$

for some β found from tables of the Gaussian distribution.

8.3. Classical and Bayesian Approaches to Confidence Intervals

You have just been introduced to confidence intervals, and have learned to make claims such as, "At the 95% level, $|\bar{x} - \mu| < 1.96\sigma/\sqrt{n}$." Let us

†Specifically, it isn't a *sufficient*[9] statistic, a concept we shall not discuss here.

now consider this statement more carefully. For simplicity let us continue to assume that our sample is drawn from an $N(\mu, \sigma^2)$ population.

We have a sample drawn from a population with mean μ; although we don't know its value μ is some definite number such as 5.02 or 19.58.[9,10] The sample mean $\bar{x}$, on the other hand, is a random variable. Our confidence interval is a statement of our determination to reject any proposed value of μ that leads to a probability of less than 5% of obtaining our value of $\bar{x}$. If we were given a large number of independent samples drawn from the same population we would expect that 95% would lie in our confidence interval.

You should note that in the last paragraph, summarising the classical theory of confidence intervals, we have *not* made statements such as, "The probability that μ lies in the interval $[\bar{x} - 1.96\sigma/\sqrt{n}, \bar{x} + 1.96\sigma/\sqrt{n}]$ is 95%." Such a remark is considered meaningless, as μ is simply a constant. If you want to make such remarks you will have to become a supporter of the Rev. Bayes.

In the Bayesian view, we have one sample with known mean $\bar{x}$, and it is μ that is the random variable.[11,12] Bayes' theorem states that

$$p(\mu|\mathbf{x}) = \frac{p(\mathbf{x}|\mu)p(\mu)}{p(\mathbf{x})},$$

where we interpret $p(\mu|\mathbf{x})$ as being the probability distribution of the population mean (it is sometimes thought of as being the distribution of how strongly we believe that the population mean has the value μ). The sample values $\mathbf{x}$ have been measured, so $p(\mathbf{x})$ is just a number. We'll adopt the Bayesian postulate for μ, and find that

$$\begin{aligned} p(\mu|\mathbf{x}) &\propto \prod_i \frac{1}{\sigma} e^{-(x_i-\mu)^2/2\sigma^2} \\ &= \sigma^{-n} e^{-n(s^2+(\bar{x}-\mu)^2)/2\sigma^2}. \end{aligned}$$

The sample variance s^2 is also just some measured number and may be absorbed into the normalization constant, and we finally arrive at

$$p(\mu|\mathbf{x}) = \frac{\sqrt{n}}{\sqrt{2\pi}\sigma} e^{-n(\bar{x}-\mu)^2/\sigma^2}.$$

We can now place confidence intervals on μ directly from its probability distribution, for example (at the 95% level)

$$|\bar{x} - \mu| < 1.96\sigma/\sqrt{n},$$

which is the same result as before, but with the difference that now it is $\bar{x}$ that is fixed, and μ the random variable.

Bayesians are not required to adopt uniform priors. For instance, if I knew that the value of a parameter lay in the range $[0, 1]$ I could choose a prior

$$p(\theta) = \begin{cases} 1 & \text{if } 0 \le \theta \le 1, \\ 0 & \text{otherwise.} \end{cases}$$

Classical statisticians are unable to use this information, as they have sworn to regard θ as an (unknown) constant and it doesn't help to know that it lies in some interval.[13]

Problem 54. A Bayesian astronomer believes that the dimensionless Hubble constant h is 80 ± 15. When she reads a paper claiming that it is in fact 60 ± 10, how should she adjust her beliefs?[14,15] Assume that both distributions are Gaussian.

Problem 55. The age of the universe t_0 is given by $1/h$ (see the previous problem, and ignore various important pieces of astrophysics); a value of $h = 100$ corresponds to 10^{10} years. Taking all due uncertainties into account, our heroine from problem 54 decides that t_0 is 10 ± 5 billion years. She then reads a book claiming that the universe was created 6000 ± 5 years ago, the uncertainty coming from the book's unknown publication date. If both distributions are taken to be Gaussian, what should she conclude about the age of the universe, from the points of view of, first, Bayesian statistics, and then common sense?

Let us now apply the Bayesian approach to the problem of setting confidence limits when the population variance σ^2 is unknown. We'll adopt the prior $p(\mu, \sigma) = 1/\sigma$ (i.e., μ's prior is still uniform, σ's prior is as usual $1/\sigma$, and μ and σ are independent),[16] in which case we see that

$$p(\mu, \sigma|\mathbf{x}) \propto \sigma^{-(n+1)} e^{-n(s^2+(\bar{x}-\mu)^2)/2\sigma^2}$$

and we can obtain $p(\mu|\mathbf{x})$ by integrating over σ:

$$p(\mu|\mathbf{x}) = \int_0^\infty p(\mu, \sigma|\mathbf{x})\, d\sigma$$

$$\propto \int_0^\infty \sigma^{-(n+1)} e^{-n(s^2+(\bar{x}-\mu)^2)/2\sigma^2}\, d\sigma.$$

Substituting $z = n(s^2 + (\bar{x}-\mu)^2)/2\sigma^2$, so $dz/z = -2d\sigma/\sigma$, and dropping a constant coefficient, this becomes

$$\left(s^2 + (\bar{x}-\mu)^2\right)^{-n/2} \int_0^\infty z^{n/2-1} e^{-z}\, dz$$

$$\propto \left(1 + \left(\frac{\bar{x}-\mu}{s}\right)^2\right)^{-n/2}.$$

Writing $n = \nu + 1$ and $t^2 = (\bar{x}-\mu)^2/(s^2/(n-1))$ we finally arrive at

$$p(\mu|\bar{x}) \propto \left(1 + t^2/\nu\right)^{-(\nu+1)/2},$$

and we have rediscovered the usual t-distribution (but again with the difference that it is μ that is considered a random variable rather than $\bar{x}$ and s^2). The presence of the "nuisance" parameter σ (i.e., a parameter in whose value we were not interested, but which appeared in our formulas) caused no trouble, beyond forcing us to evaluate an extra integral.

References

1: *K&S* 20.3
2: *K&S* 20.5
3: *K&S* 20.31
4: *NR* 13.4
5: *K&S* 20.8
6: *K&S* 20.16
7: *Lee* 2.6
8: *K&S* 20.14
9: *K&S* 20.4
10: *Lee* 2.6
11: *K&S* 21.28
12: *Lee* 2.6
13: *K&S* 21.36
14: *Lee* 2.2
15: *K&S*$_V$ 8.15
16: *Lee* 2.12

9. Hypothesis Testing

If we want to answer a question such as "Given a sample drawn from a Gaussian distribution, is the population mean μ_0?" how should we proceed? We can calculate $\bar{x}$ from the sample and use the methods discussed in the previous section to give the probability that, if the population mean were really μ_0, we would have found a given value of $\bar{x}$, or we can say that μ probably lies in such-and-such an interval; but these don't quite answer the question.

First we'll replace the original question by a *null hypothesis*[1] (usually written H_0); in this case H_0 would be "The population mean is μ_0." The original question states that the parent distribution is Gaussian, so there is no need to include the claim in H_0; if we were intending to test that too we'd have chosen a null hypothesis such as "The parent distribution is Gaussian and the population mean is μ_0." We can then restate the original question as "Is H_0 true?" There are two different wrong ways to answer this: we could decide that H_0 is false whereas it is in fact true, or we could accept H_0 when it is in fact false. These mistakes are called *type I* and *type II* errors, respectively; they are also sometimes called "false negatives" and "false positives."[2] The difficulty is that when considering type II errors we don't know the properties of the population from which our sample was drawn, so we can't calculate the probability of a type II error without specifying the alternative that we have in mind. For example, the alternative (call it H_1) might be "The population mean is greater than μ_0," or it might be "The population mean could have any value."

Hypotheses may be divided into *simple* or *composite*;[3] simple hypotheses are fully specified, e.g., "The population is $N(\mu, \sigma^2)$," whereas composite ones involve unknown parameters, e.g., "The population is Gaussian, with unknown mean and variance."

In order to solve a number of common problems there is no need to delve further into the theory of hypothesis testing, as we can use intuition to invent suitable tests. For example, if H_0 is "The population mean is μ_0,"[4] and we know that the parent population is Gaussian with variance σ^2, we can simply construct a confidence interval for $\bar{x}$ assuming that H_0 is true (e.g., $\mu_0 \pm 1.96\sigma/\sqrt{n}$ at the 95% level), and reject H_0 if the sample mean does not fall within it. The range of values of $\bar{x}$ that lead to the rejection of H_0 is known as the *critical region* for the test.[5] This use of confidence intervals is slightly different, as previously we placed limits on the values of the parent population's parameters, whereas here we are assuming that we know them (by assuming explicitly that H_0 holds) and placing a limit upon the acceptable properties of the sample.

Problem 56. Construct a test for H_0 if we do not know the variance of the population.

We may have appeared to have been ignoring H_1 while constructing tests, but in fact we weren't. We chose a 2-sided confidence interval because we had some H_1 such as "The mean is not μ_0" in mind. Had we had an H_1 such as "The mean is less than μ_0" we would have been led to use a 1-sided test, based upon a 1-sided confidence interval.

We can construct similarly intuitive tests for a number of popular problems:

9.1. Do Two Samples Have the Same Mean?

We are about to discuss the more popular tests based on Gaussian parents, but you should be aware that if you don't want to make assumptions about the parent populations you can use a Wilcoxon test, which we shall discuss when we come to non-parametric statistics (section 15.3).

If we have two samples of sizes n_x and n_y, both drawn from Gaussian populations with the same standard deviation σ, how do we test the hypothesis "$\mu_x = \mu_y$"?[6,7,8] This is very similar to the problem of testing "$\mu = \mu_0$," as it may be shown that

$$\frac{\bar{x} - \bar{y}}{\sqrt{\frac{n_x s_x^2 + n_y s_y^2}{n_x + n_y - 2}\left(\frac{1}{n_x} + \frac{1}{n_y}\right)}}$$

follows a t-distribution with $n_x + n_y - 2$ degrees of freedom, so we can use confidence intervals based upon this statistic to test our hypothesis. Although the derivation of this result depends on the Gaussian nature of the parent populations, it is in fact quite robust to departures from normality.[9] In section 15.3 we shall meet a test (the normal-scores variant of the Wilcoxon test) that is always at least as good as this t-test and which makes no assumptions about the nature of the parent population.

Problem 57. Consider two samples **x** and **y** drawn from parent populations $N(\mu_x, \sigma_x^2)$ and $N(\mu_y, \sigma_y^2)$. What is the distribution of $\bar{x} - \bar{y}$? Show that

$$\frac{\sum(x_i - \bar{x})^2}{\sigma_x^2} + \frac{\sum(y_i - \bar{y})^2}{\sigma_y^2}$$

follows a $\chi^2_{n_x+n_y-2}$ distribution. If $\sigma_x = \sigma_y \equiv \sigma^2$, deduce that if I define

$$s^2 = \frac{n_x s_x^2 + n_y s_y^2}{n_x + n_y - 2}$$

then $\langle s^2 \rangle = \sigma^2$. For a Gaussian population we know that $\langle \bar{x} s_x^2 \rangle = 0$. Show that s^2 and $\bar{x} - \bar{y}$ are independent and hence that

$$\frac{\bar{x} - \bar{y}}{\sqrt{\frac{n_x s_x^2 + n_y s_y^2}{n_x + n_y - 2}\left(\frac{1}{n_x} + \frac{1}{n_y}\right)}}$$

follows a t-distribution with $n_x + n_y - 2$ degrees of freedom.

If the variances of the two populations are different (see section 9.3 for a way to test this) it is harder to construct a test.[10,11,12] It's easy to show that the variance of $\bar{x} - \bar{y}$ can be estimated as

$$S^2 \equiv \frac{s_x^2}{n_x - 1} + \frac{s_y^2}{n_y - 1}$$

and that S and $\bar{x} - \bar{y}$ are independent. Unfortunately S^2 is not a multiple of a χ^2 distribution unless $\sigma_x = \sigma_y$, and therefore $(\bar{x} - \bar{y})/S$ doesn't follow a t-distribution.

It turns out that we can still carry out a test based on

$$\frac{\bar{x} - \bar{y}}{S} = \frac{\bar{x} - \bar{y}}{\sqrt{\frac{s_x^2}{n_x - 1} + \frac{s_y^2}{n_y - 1}}},$$

as it is distributed approximately as t with[13]

$$\frac{\left(\frac{s_x^2}{n_x - 1} + \frac{s_y^2}{n_y - 1}\right)^2}{\frac{(s_x^2/n_x - 1)^2}{n_x - 1} + \frac{(s_y^2/n_y - 1)^2}{n_y - 1}}$$

degrees of freedom.

Problem 58. Calculate S^2's mean and variance. If we want to approximate S^2 as a multiple of a χ^2_ν variable, what value of ν should we choose? Deduce that $(\bar{x} - \bar{y})/S$ is approximately a t_ν variate. (*Hint:* what are the mean and variance of a χ^2_α variate?)

You shouldn't be concerned that ν isn't an integer; you can either interpolate in tables of t, or use beta functions and not worry about it.[14] Be warned that if the distributions have different variances they may have different shapes too, which means that studying the difference in their means may not be very useful.

The problem of two means is one of the classics of statistics,[15,16] so let us discuss it from a Bayesian viewpoint; our conclusions will not change much.

We can write down $p(\mu_x, \mu_y, \sigma_x, \sigma_y; \mathbf{x}, \mathbf{y})$ by direct analogy to our discussion of the Bayesian theory of sampling, and integrate it over σ_x and σ_y to obtain $p(\mu_x, \mu_y; \mathbf{x}, \mathbf{y})$. The resulting distribution is the product of two independent t-distributions, $t^x_{\nu_x}$ and $t^y_{\nu_x}$, where

$$t^x_{\nu_x} = \frac{\bar{x} - \mu}{s_x / \sqrt{\nu_x}},$$

$n_x = \nu_x + 1$, and analogous definitions hold for n_y and $t^y_{\nu_y}$. If we define

$$\tan^2\theta = \frac{s_x^2/\nu_x}{s_y^2/\nu_y},$$

then

$$z \equiv \frac{\bar{x} - \bar{y}}{S} = t^x_{\nu_x}\sin\theta - t^y_{\nu_y}\cos\theta.$$

Because θ is a constant (as we know s_x^2 and s_y^2) we can calculate z's distribution; it is usually called the *Behrens-Fisher distribution* $B_F(\nu_x, \nu_y, \theta)$ and has been tabulated.[17,18] If

$$f_1 = \frac{\nu_x - 1}{\nu_x - 3}\sin^2\theta + \frac{\nu_y - 1}{\nu_y - 3}\cos^2\theta,$$

and

$$f_2 = \frac{(\nu_x - 1)^2}{(\nu_x - 3)^2(\nu_x - 5)}\sin^4\theta + \frac{(\nu_y - 1)^2}{(\nu_y - 3)^2(\nu_y - 5)}\cos^4\theta$$

then z/α is approximately a t_β variable with

$$\beta = 4 + f_1^2/f_2$$

and

$$\alpha^2 = f_1(\beta - 2)/\beta.$$

For large ν_x and ν_y these reduce to $\alpha = 1$ and $\beta = \nu$, where ν is the complicated expression given above.[19]

You should note that the Behrens-Fisher distribution came about because θ is a known constant from a Bayesian perspective, whereas classical theory would need to estimate σ_x/σ_y from the data. In fact classical theory is unable to give a satisfactory answer to the question "What is the distribution of z?" as it doesn't know the value of θ. This may not be too important in practice, as the approximate distribution, which results from using the data to estimate θ, agrees with our Bayesian result.

9.2. Comparing the Means of Paired Samples

Sometimes a two-sample problem turns out to be a one-sample problem in disguise; an example is testing the means of two samples where each member of the first sample corresponds to a member of the second.[20,21,22] For instance, suppose I want to know if a class has benefited from a lecture course that I have given. I could test the hypothesis that the mean score on some test has risen, but this would probably not give me a significant answer, as the spread in scores (e.g., the variance) is so much larger than the gain (or loss) resulting from my endeavours. What I should do instead is concentrate on the change in individual scores; i.e., if the scores before and after are x and y, I should consider the distribution of $\Delta_i = x_i - y_i$ (where the subscript refers to the same student in the x and y samples). The Δ's are independent variables, and if x and y are Gaussian so is Δ, so I can apply a standard t-test to test the hypothesis "$\bar{\Delta} = 0$." If you want a formula, it is clear that

$$s_\Delta^2 = s_x^2 + s_y^2 - 2s_{xy},$$

so

$$\frac{\bar{x} - \bar{y}}{\sqrt{\frac{s_x^2 + s_y^2 - 2s_{xy}}{n-1}}}$$

follows a t-distribution with $n - 1$ degrees of freedom.

9.3. Do Two Samples Have the Same Variance?

If we have two samples, both drawn from Gaussian populations, we can test if $\sigma_x^2 = \sigma_y^2$ by considering the ratio s_x^2/s_y^2. We'd expect it to be close to 1 if the parent variances are equal, and if only we knew its distribution we'd be able to set up confidence intervals and test hypotheses.[23,24]

We know that $\sum(x_i - \bar{x})^2/\sigma^2$ is a $\chi^2_{n_x-1}$ variate, so the ratio

$$f = \frac{\sum(x_i - \bar{x})^2/(n_x - 1)}{\sum(y_i - \bar{y})^2/(n_y - 1)}$$

is of the form

$$\frac{\chi^2_\alpha/\alpha}{\chi^2_\beta/\beta},$$

which we know to follow the $F_{\alpha,\beta}$-distribution; our variance ratio therefore follows an F_{n_x-1,n_y-1}-distribution, which is a well-tabulated function. Because we are usually interested in testing if the two variances are equal, we want to reject both large and small values of f; i.e., we usually use a 2-sided test. It is obvious (as well as provable from the properties of incomplete beta functions[25]) that

$$\int_0^{1/x} dF = \int_x^\infty dF,$$

i.e., that the two tails are equal.

You should be warned that this test is sensitive to departures of the parent populations from normality; after all, it depends upon the distribution of s^2, which depends on all of the moments of the parent population up to and including the fourth.[26,27]

9.4. The Theory of Tests

Let us return to a formal consideration of the problem of choosing tests, and start by considering type I errors: choose some region $\mathcal{R}$ in n-dimensional sample space so that

$$\int_{\mathcal{R}} f\, d^n x = 1 - \alpha,$$

i.e., the probability of a type I error is α. Now what about type II errors? We have seen that we can choose the interval (for a given α) in any number of ways; let us choose the particular one that minimizes the probability of a type II error (one minus this probability is called the *power* of the test).[28] This is the fundamental idea introduced by Neyman and Pearson that underlies the theory of hypothesis testing. The power will usually depend upon our choice of H_1.

Let us adopt H_0 : mean is μ_0 and H_1 : mean is μ_1. Let us further consider a sample of size 2, and draw a graph of x_1 against x_2 so the sample will be represented by a certain dot (see the figure).[29] We can also draw the contour lines corresponding to the x_i's p.d.f. (assuming that H_0 is true); they will be symmetrical about the line $x_1 = x_2$ with "centre of gravity" $(x_1, x_2) = (\mu_0, \mu_0)$. I can choose any region to be my critical region for rejecting H_0, provided that the integral of the p.d.f. over the region is α.

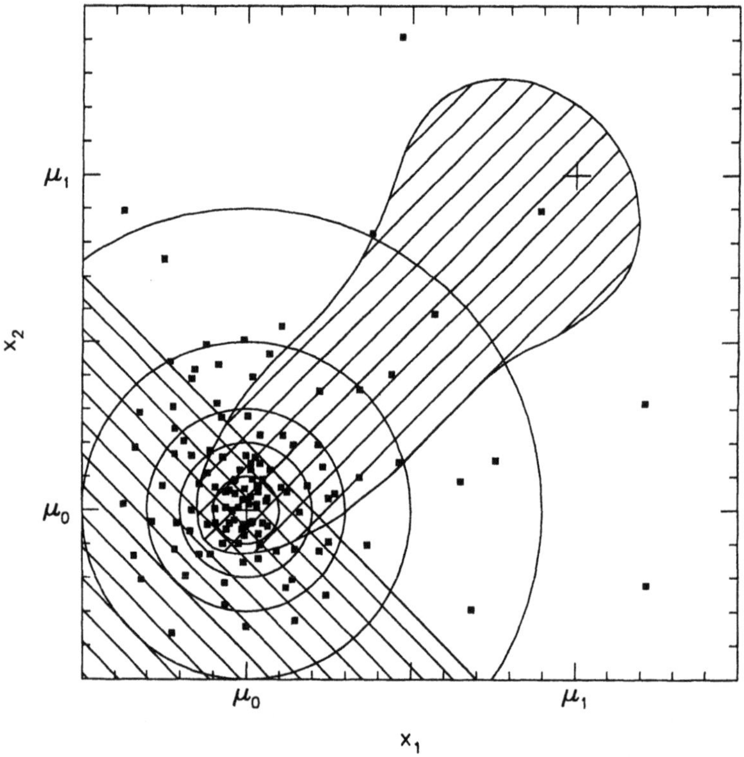

The dots represent different samples, while the circles represent the p.d.f. when H_0 is true. The two crosses represent the points (μ_0, μ_0) and (μ_1, μ_1), and the two shaded regions illustrate two possible confidence regions.

The two shaded regions in the figure represent two possible confidence regions, as the integral of the p.d.f. over either of them is $1 - \alpha$ (when H_0 holds).

We have decided to choose between possible regions by minimizing the probability of a type II error, so the best choice of a region is the one that excludes as much of the second p.d.f. as possible. When we assume that H_1 is true and consider the two regions in the figure, we see that the pear-shaped region has the higher probability of a type II error (imagine integrating the p.d.f. over it; it encloses the mean of the distribution so its integral must surely be larger than for the region in the lower left); in this case we should choose the lower-left-hand region.

If we write $L(x_i|H)$ as the probability of a given set of x_i if H is true (L is called the likelihood, and we'll be seeing lots more of it soon), the boundary of the acceptance region is given by[30]

$$\frac{L(x_i|H_0)}{L(x_i|H_1)} = k_\alpha,$$

where k_α depends on the value of α chosen; this is known as the *Neyman-Pearson lemma.*†If this were not true we could make the probability of a type II error smaller by adjusting the boundary to enclose regions of larger $L(H_0)/L(H_1)$ and exclude regions of smaller $L(H_0)/L(H_1)$ while keeping α constant. In general the boundary will depend upon H_1; in some cases the boundary *doesn't* depend upon H_1 (or in this case, upon μ_1) and the critical region is called *uniformly most powerful* (UMP).[31] Once the boundary of the critical region has been found we can proceed to determine the power of the test, noting that there is no guarantee that an efficient estimator will give rise to a powerful test.

If x_1 and x_2 are drawn from an $N(\mu, \sigma^2)$ distribution, it's easy to show that we should choose a region whose boundary is perpendicular to the line joining the origin to (μ_1, μ_1), and it's clear that the position of the line is independent of the value of μ_1[32] and that we can accept or reject H_0 based solely upon the value of $\bar{x}$ rather than being forced to consider all of the x_i; the region is a UMP one.

Problem 59. Suppose we know that the parent distribu-

†This is E. S. Pearson, the son of K. Pearson who invented the χ^2 test. He was married to one of my great aunts.

tion of a two-element sample is a Cauchy distribution:

$$dF = \frac{1}{\pi} \frac{1}{1 + (x - \mu)^2} \, dx.$$

What confidence region should we use to decide between the hypotheses $H_0 : \mu = 0.5$ and $H_1 : \mu = 1.5$?

Even for the problem of determining whether the mean of a Gaussian population is μ_0 there is in general no uniformly most powerful region if we don't know whether μ_1 is larger or smaller than μ_0. We may be tempted to hedge our bets and choose to use the two boundaries corresponding to $\mu_1 < \mu_0$ and $\mu_1 > \mu_0$ each with a type I risk of $\alpha/2$, and although this is hard to justify theoretically it is widely done in practice.[33]

Problem 60. If I adopt the alternative hypothesis H_1: "I don't know anything about the parent distribution," what is $L(x_i|H_1)$? What form does the Neyman-Pearson lemma now take? For our standard problem of a Gaussian parent distribution of known variance and an H_0 of "The population mean is μ_0," what should I choose for my region $\mathcal{R}$? If I decide to focus attention not on the distribution of n x_i but on the single variable $\bar{x}$, what is $L(\bar{x}|H_0)$, and what region should I adopt to test H_0?

The Bayesians, of course, have their own approach to hypothesis testing. They calculate the posterior probabilities $p_0 = \Pr(\mathbf{x}|H_0)$ and $p_1 = \Pr(\mathbf{x}|H_1)$ and adopt the hypothesis with the higher probability.[34] In general their conclusions differ from those of the classicists, but one case in which both sides agree is that of one-sided tests:

Suppose that I have a sample of n points drawn from an $N(\mu, 1)$ population, so my sample mean $\bar{x}$ has variance $1/n$, and that I want to test $H_0 : \mu < \mu_0$ against $H_1 : \mu \geq \mu_0$. The classical solution would be to say that I can reject H_0 at a level p_0 given by

$$1 - p_0 = \sqrt{\frac{n}{2\pi}} \int_{\bar{x}}^{\infty} e^{-n(t-\mu_0)^2/2} \, dt;$$

that is, the probability of getting a value at least as large as $\bar{x}$ is $1 - p_0$.

If I now don my Bayesian robes, then adopt a uniform prior for μ, I find that μ's posterior probability distribution is $N(\bar{x}, 1/n)$. I can then proceed to calculate the probability that H_0 is true:

$$p_0 = 1 - p_1 = \sqrt{\frac{n}{2\pi}} \int_{\mu_0}^{\infty} e^{-n(t-\bar{x})^2/2}\, dt,$$

which is seen to agree with the classical result.

Although the results in this case are the same, the underlying ideas are very different. A Bayesian would describe a classical statistician as first predicting that we will observe no values greater than $\bar{x}$ if H_0 is true, and then being gratified when the unexpected values do not, in fact, materialize.

9.5. Likelihood Ratio Tests

The Neyman-Pearson lemma has directed our attention to the likelihood, and it turns out that we can use L to construct tests.[35] We shall use such a test to estimate the significance of parameters in the context of linear models, but as we shall also derive the result by elementary methods, feel free to ignore this section.

Consider the hypothesis $H_0 : \theta_r = \theta_{r0}$ with $H_1 : \theta_r \neq \theta_{r0}$. It is possible to construct tests based directly upon the likelihood $L(\mathbf{x}|\theta_r, \theta_s)$ (θ_s are the parameters not fixed by H_0) as follows:

First find the absolute maximum likelihood, unconstrained by H_0, $L(\mathbf{x}|\hat{\theta}_r, \hat{\theta}_s)$, and then find the maximum when H_0 holds, $L(\mathbf{x}|\theta_{r0}, \hat{\hat{\theta}}_s)$ (note that $\hat{\hat{\theta}}_s$ will not in general equal $\hat{\theta}_s$). Now form the ratio

$$\ell = \frac{L(\mathbf{x}|\theta_{r0}, \hat{\hat{\theta}}_s)}{L(\mathbf{x}|\hat{\theta}_r, \hat{\theta}_s)},$$

which clearly lies in the interval $[0, 1]$. We should accept H_0 if ℓ is large, so the critical region is

$$\ell \le c_\alpha$$

for a test of size α. For obvious reasons such tests are known as *likelihood ratio* (LR) *tests.*

Let's consider our standard example, a sample of size n drawn from an $N(\mu, \sigma^2)$ distribution with the null hypothesis $H_0 : \mu = \mu_0$. The likelihood function is[36]

$$L(\mu, \sigma^2) = \frac{1}{(2\pi)^{n/2}\sigma^n} \exp(-\sum(x_i - \mu)^2/2\sigma^2),$$

and, as we saw,

$$\hat{\mu} = \bar{x}$$

and

$$\hat{\sigma}^2 = s^2 \equiv \frac{1}{n} \sum (x_i - \bar{x})^2,$$

so

$$L(\hat{\mu}, \hat{\sigma}^2) = (2\pi s^2)^{-n/2} e^{-n/2}.$$

When H_0 holds,

$$\begin{aligned} \hat{\hat{\sigma}}^2 &= \frac{1}{n} \sum (x_i - \mu_0)^2 \\ &= s^2 + (\bar{x} - \mu_0)^2 \end{aligned}$$

and

$$L(\mu_0, \hat{\hat{\sigma}}^2) = (2\pi (s^2 + (\bar{x} - \mu_0)^2))^{-n/2} e^{-n/2},$$

so

$$\begin{aligned} \ell &= \left(\frac{s^2}{s^2 + (\bar{x} - \mu_0)^2} \right)^{n/2} \\ &= \left(1 + t^2/(n-1) \right)^{-n/2}, \end{aligned}$$

where t follows Student's t-distribution with $n - 1$ degrees of freedom. Because ℓ is a monotonic function of a known distribution it's easy to find the critical value c_α from tables of t; the critical region is 1-tailed in t^2 and thus 2-tailed in t.

In general it isn't easy to find the distribution of ℓ, and various approximation schemes have been developed.[37] It is possible to show that asymptotically (as the MLE estimators tend to normality) the distribution of $-2 \ln \ell$ tends to a χ^2_ν distribution with ν equal to the number of parameters in the model. In this limit the likelihood ratio test tends to a test based upon the MLE estimators themselves.[38] In order to estimate the power of the test (the probability of not making a type II error) we would have to study the theory of non-central χ^2 distributions, a pleasure that we shall forego.

References

1: *K&S* 22.6
2: *K&S* 22.8
3: *K&S* 22.4
4: *Sachs* 3.11, 3.2
5: *K&S* 22.5
6: *K&S* 21.12
7: *Sachs* 3.6.1
8: *Lee* 5.1
9: *K&S* 31.4
10: *K&S* 21.13
11: *Lee* 5.4
12: *Sachs* 3.6.2
13: *K&S* 21.24
14: *NR* 13.4
15: *K&S* 21.31
16: *Lee* 5.3
17: *K&S* 21.26
18: *K&S*$_V$ 20.44
19: *Patil*
20: *K&S* 31.76
21: *Lee* 5.1
22: *Sachs* 4.2.1
23: *Lee* 5.5
24: *Sachs* 3.5
25: *K&S* 16.20
26: *K&S* 31.6
27: *Sachs* 3.5.1
28: *K&S* 22.8
29: *K&S* Ex. 22.1
30: *K&S* 22.10
31: *K&S* 22.16
32: *K&S* Ex. 22.2
33: *K&S* 22.26
34: *Lee* 4.1
35: *K&S* 24.1
36: *K&S* 24.2
37: *K&S* 24.3, 24.7
38: *K&S* 24.7

10. The Theory of Maximum Likelihood Estimators

We have already discussed Bayes' theorem and used maximum likelihood as a way of estimating the properties of unknown parameters, constructing confidence intervals, and testing hypotheses. Let us now turn to some more formal properties of ML estimators.

10.1. Efficiency of ML Estimators and the Minimum Variance Bound

It is clear that[1]

$$\int L(x_i; \theta)\, d^n x = 1$$

(θ is a parameter to be estimated) and therefore, differentiating with respect to θ, that

$$\int \frac{\partial L}{\partial \theta}\, d^n x = 0,$$

i.e.,

$$\int \frac{1}{L}\frac{\partial L}{\partial \theta} L\, d^n x \equiv \left\langle \frac{\partial \ln L}{\partial \theta} \right\rangle = 0.$$

Differentiating again,

$$\int \frac{\partial \ln L}{\partial \theta}\frac{\partial L}{\partial \theta} + \frac{\partial^2 \ln L}{\partial \theta^2} L\, d^n x = 0,$$

i.e.,

$$\left\langle \left(\frac{\partial \ln L}{\partial \theta} \right)^2 \right\rangle = -\left\langle \frac{\partial^2 \ln L}{\partial \theta^2} \right\rangle .$$

Now consider some estimator t, which is unbiased for some function of θ, $\tau(\theta)$:[2]

$$\tau(\theta) = \langle t \rangle \equiv \int t L\, d^n x,$$

so

$$\frac{d\tau}{d\theta} \equiv \tau'(\theta) = \int \frac{t}{L}\frac{\partial L}{\partial \theta} L\, d^n x.$$

Noting that

$$\int \tau \frac{\partial \ln L}{\partial \theta} L\, d^n x = \tau \int \frac{\partial \ln L}{\partial \theta} L\, d^n x = \tau \left\langle \frac{\partial \ln L}{\partial \theta} \right\rangle = 0,$$

we have

$$\tau' = \int (t - \tau) \frac{\partial \ln L}{\partial \theta} L\, d^n x$$

and, remembering the Cauchy Inequality (that $(f \cdot g)^2 \le f^2 g^2$, where the inner product $f \cdot g \equiv \int fgL\, d^n x$), we can see that

$$\tau'^2 \le \int (t-\tau)^2 L\, d^n x \times \int \left(\frac{\partial \ln L}{\partial \theta}\right)^2 L\, d^n x,$$

i.e.,

$$\tau'^2 \le V(t) \times \left\langle \left(\frac{\partial \ln L}{\partial \theta}\right)^2 \right\rangle,$$

which is called the *Cramér-Rao Inequality* and sets a lower limit on the variance of an estimator. For the case $\tau(\theta) = \theta$, the inequality becomes

$$V(t) \ge \frac{1}{\left\langle \left(\frac{\partial \ln L}{\partial \theta}\right)^2 \right\rangle} = \frac{-1}{\left\langle \frac{\partial^2 \ln L}{\partial \theta^2} \right\rangle}.$$

This is known as the *minimum variance bound* (MVB). We cannot hope to find an unbiased estimator more efficient than one that attains the MVB.

To attain the MVB we require that[3]

$$t - \tau \propto \frac{\partial \ln L}{\partial \theta},$$

i.e.,

$$A(\theta)(t-\tau) = \frac{\partial \ln L}{\partial \theta}$$

(as $(f \cdot g)^2 = f^2 g^2$ if and only if f and g are "parallel"). Multiplying by $(t-\tau)$ and taking expectation values,

$$\left\langle A(\theta)(t-\tau)^2 \right\rangle = \left\langle (t-\tau)\frac{\partial \ln L}{\partial \theta} \right\rangle = \tau',$$

so

$$V(t) = \frac{\tau'}{A(\theta)},$$

or, if $\tau(\theta) = \theta$,

$$V(t) = \frac{1}{A(\theta)}.$$

Returning to our Gaussian example, when estimating the mean we found that[4]

$$\frac{\partial \ln L}{\partial \mu} = \frac{1}{\sigma^2}\sum_i (x_i - \mu).$$

The parameter to be fit is called μ rather than θ, and it's clear that if we choose $t = \bar{x}$ and $\tau(\theta) = \theta$ we have satisfied the conditions required to attain the MVB: $A = n/\sigma^2$ and the variance is σ^2/n. This is the result that we referred to while discussing the efficiency of the median as an estimator of the mean of a Gaussian.

Now consider trying to estimate σ (i.e., $\theta = \sigma$). The likelihood is maximized when

$$\frac{\partial \ln L}{\partial \sigma} = -\frac{n}{\sigma^3}\left(\sigma^2 - \frac{1}{n}\sum_i (x_i - \mu)^2\right),$$

which is not in the correct form to deduce an MVB estimator for σ. On the other hand it *is* in the right form for $\sum(x - \mu)^2/n$ to be an MVB estimator for σ^2, and setting $\tau(\theta) = \theta^2$ and $A = n/\sigma^3$ we find that the MVB is $2\sigma^4/n$.

In general, ML estimators don't attain the MVB, but it is possible to prove that asymptotically (i.e., as $n \to \infty$) they are unbiased, consistent, and do attain the MVB.[5] They are also asymptotically Gaussian (the proof follows from the fact that $\ln L$ consists of the sum of a large number of independent terms so the central limit theorem applies).[†]It is the lure of these optimal asymptotic properties that makes MLE so popular, although, as we saw in problem 48, ML can lead to surprising conclusions. Unfortunately in the real world sample sizes are seldom large enough that the promised asymptotic land is reached; in particular ML estimators are often both biased and non-Gaussian.

Problem 61. In a study of the dynamics of the globular star cluster M13,[6] Jim Gunn, Roger Griffin, and I measured radial velocities for about 130 stars, with the intention of measuring the cluster's velocity dispersion. Each star had a measured velocity v_i and error Δ_i; we also made a model which predicted that the velocity of the ith star should be $\bar{v} + v_i$ where the mean cluster velocity is $\bar{v}$ and the random velocity of the star is v_i. The model assumed that the velocity dispersion at each point in the cluster is Maxwellian (so the 1-dimensional distribution is Gaussian) with 1-dimension dispersion $v_0\eta_i$ (the projected

---

[†]Because a Gaussian is symmetrical about its mode, the choice of the maximum rather than the median or mode becomes moot in the large n limit.

velocity dispersion drops off with distance from the centre of the cluster, which is why the dimensionless velocity dispersion η_i requires a subscript). The values of $\bar{v}$ and v_0 are to be found from the data. Write down the likelihood function and derive maximum likelihood equations for v_0 and $\bar{v}$. How would you go about solving these if $\Delta_i \ll v_o \eta_i$?

10.2. Example: Statistical Parallaxes by MLE

Consider the problem of estimating the distance to a group of stars, all known to be at the same distance (e.g., a star cluster, or a set of Cepheid variables corrected for distance). Assume that the velocity ellipsoid is isotropic and that the distribution function is Maxwellian with velocity dispersion σ^2:

$$dF(\mathbf{v}) = \frac{1}{(2\pi)^{3/2}\sigma^3} e^{-\mathbf{v}^2/2\sigma^2} d^3\mathbf{v}.$$

The true velocity of the ith star is given by

$$\mathbf{v}_i = \Lambda d \boldsymbol{\mu}_i + \mathbf{v}_{r,i},$$

where Λ is a constant (converting between radians, parsecs, kilometres, seconds, and years), d the distance to the stars, $\boldsymbol{\mu}$ the proper motion,† and $\mathbf{v}_r$ the radial velocity. By a clever choice of units we can make $\Lambda = 1$; let us assume that this has been done.

The likelihood for the ith star is given by

$$\begin{aligned} L_i &= P_i(\mathbf{v}_i | d, \sigma) \\ &= \frac{1}{(2\pi)^{3/2}\sigma^3} e^{-\mathbf{v}_i^2/2\sigma^2}, \end{aligned}$$

but we have measured v_r and $\boldsymbol{\mu}$ rather than $\mathbf{v}$, so, changing variables to $(v_r, \boldsymbol{\mu})$, the likelihood becomes

$$\begin{aligned} L_i &= P_i(v_{r,i}, \boldsymbol{\mu}_i | d, \sigma) \\ &= \frac{d^2}{(2\pi)^{3/2}\sigma^3} e^{-(v_{r,i}^2 + \mu_i^2 d^2)/2\sigma^2}. \end{aligned}$$

†That is, the apparent velocity of the star across the sky.

The log-likelihood of a given sample of size n is then

$$2n \ln d - \frac{3n}{2} \ln(2\pi) - 3n \ln \sigma - \frac{1}{2\sigma^2} \sum_i (v_{r,i}^2 + \mu_i^2 d^2),$$

and maximizing with respect to σ and d we find that

$$\frac{3}{\sigma} = \frac{1}{\sigma^3}\left(\left\langle v_r^2 \right\rangle + d^2 \left\langle \mu^2 \right\rangle\right)$$

and

$$\frac{2}{d} = \frac{d}{\sigma^2} \left\langle \mu^2 \right\rangle,$$

respectively, so

$$\hat{\sigma}^2 = \left\langle v_r^2 \right\rangle$$

and

$$\hat{d}^2 = \frac{2\sigma^2}{\langle \mu^2 \rangle}.$$

Remembering that

$$\mu^2 = \mu_x^2 + \mu_y^2,$$

these are eminently reasonable.

In this analysis we knew the form of the likelihood directly from the model, but sometimes things are not quite so simple and we must calculate the distribution. For example, an alternative treatment of the same problem writes

$$\boldsymbol{\eta}_i = \Lambda b \boldsymbol{\mu}_i + \mathbf{v}_{r,i},$$

where b is now an assumed distance, $b = (1 + k)d$. If we write the projection operator (onto the plane of the sky) for the ith star as $P_i \equiv I - \mathbf{r}_i \mathbf{r}_i^T = P^T$ we can write

$$\boldsymbol{\eta}_i = (I + kP_i)\mathbf{v}_i$$

and calculate its dispersion matrix:

$$\begin{aligned} M_i = \left\langle \boldsymbol{\eta}_i \boldsymbol{\eta}_i^T \right\rangle &= (I + kP_i) \left\langle \mathbf{v}_i \mathbf{v}_i^T \right\rangle (I + kP_i) \\ &= \sigma^2 (I + kP_i)(I + kP_i) \\ &= \sigma^2 (I + k(2 + k)P_i). \end{aligned}$$

The inverse may easily be shown to be

$$M_i^{-1} = \frac{1}{\sigma^2}\left(I - \frac{k(2 + k)}{(1 + k)^2} P_i\right)$$

with determinant $\sigma^{-6}(1 + k)^{-4}$.

Problem 62. Find the inverse of the matrix $I + k(2+k)P$.

We can now write down the maximum likelihood estimators:

$$L_i = \frac{1}{(2\pi)^{3/2}\sigma^3(1+k)^2}\exp(-\mathbf{v}_i^T\mathbf{v}_i + k(2+k)/(1+k)^2\mathbf{v}_i^T P_i \mathbf{v})/2\sigma^2;$$

form their product for all stars; take the log; differentiate with respect to k and σ; and rearrange to give

$$\hat{\sigma}^2 = \left\langle v_r^2 \right\rangle$$

$$\frac{b^2}{(1+\hat{k})^2} = 2\frac{\langle v_r^2\rangle}{\langle \mu^2\rangle},$$

which agrees with our previous result.

10.3. Maximum Likelihood, Chi-Squared, Least Squares, and All That

Let us imagine that we want to fit a model to some data. We have measured a set of $\{x_i\}$ (without error), and also a set of corresponding $\{y_i\}$. Each y_i has an associated error δ_i, which is drawn from a $N(0,\sigma_i^2)$ population. I have a model which predicts that $y_i = f(x_i;\boldsymbol{\theta})$ (e.g., $y_i = \theta_0 + \theta_1 x_i$); how should I estimate $\boldsymbol{\theta}$? Naturally, I choose to use maximum likelihood, and write down $\ln L$:

$$\ln L = -\frac{n}{2}\ln(2\pi) - \sum_i \ln\sigma_i - \sum_i \frac{(y_i - f(x_i;\boldsymbol{\theta}))^2}{2\sigma_i^2}.$$

Because the σ_i are known, all that I have to do to maximize this is to minimize[7]

$$X^2 = \sum_i \frac{(y_i - f(x_i;\boldsymbol{\theta}))^2}{\sigma_i^2},$$

which is simply the usual least-squares procedure.[8,9]

Can we use the value of X^2 to say something about how well the model fits the data? If we hadn't estimated any parameters from the data, then if $y_i = f(x_i;\boldsymbol{\theta}) + \delta_i$ (i.e., if the model were correct) each of the $(y_i - f(x_i;\boldsymbol{\theta}))/\sigma_i$ would have been an independent normalized Gaussian variable, X^2 would have followed the χ_n^2 distribution, and we could have used the known properties of χ^2 distributions to see whether the fit was acceptable; for example, we would have expected X^2 to lie within a few

$\sqrt{2n}$ of n. In reality we have used the data to estimate the k parameters of the model, so $f(x_i;\boldsymbol{\theta})$ and δ_j are not independent. If the model is linear in the θ_i we shall shortly (section 11.4) show that X^2 is distributed as χ^2_{n-k}, so we can still use the properties of χ^2 to test how well our model fits; otherwise we must resort to the methods of the next section.

10.4. Uncertainties in Parameters

It is not sufficient to estimate parameters and to know that (at least asymptotically) they attain the MVB. When making models and fitting parameters it is essential to provide practical estimates of their uncertainties.

If we knew the true parameters $\boldsymbol{\theta}$ of our model, and if the model were a good description of our data, then the range of values of $\hat{\boldsymbol{\theta}}$ produced by different realizations of our data would be our confidence interval[10] (see section 8.3).†

We don't know the true value of $\boldsymbol{\theta}$, but let us write $\boldsymbol{\Theta}$ as our guess and start by assuming that $\boldsymbol{\Theta} = \hat{\boldsymbol{\theta}}$. There is no guarantee that this is correct, as $\hat{\boldsymbol{\theta}}$ may be a biased estimator for $\boldsymbol{\theta}$, or we might have been unlucky and $\hat{\boldsymbol{\theta}}$ might happen to be a very poor estimate, but we can make no better an initial guess.

We want to find a volume $\mathcal{R}$ with the property that

$$\int_{\mathcal{R}} L(\hat{\boldsymbol{\theta}}|\boldsymbol{\Theta})\, d\hat{\boldsymbol{\theta}} = 1 - \alpha,$$

so that we can say (with probability $1-\alpha$ of being correct) that for any dataset we were presented with, $\hat{\theta}$ would lie in $\mathcal{R}$. This is an obvious generalization of the 1-dimensional concept of a confidence region. In order to choose $\mathcal{R}$, let's adopt the H_1: "all models are equally likely" and use the Neyman-Pearson lemma to choose a surface of constant likelihood as the boundary of our test region $\mathcal{R}$ (cf. problem 60).

If the model is linear in its parameters‡and the errors are known to be Gaussian, then it is possible to use the properties of χ^2 distributions

---

†If you want to make statements about the true value of $\boldsymbol{\theta}$ you will (as usual) have to adopt a Bayesian approach and interpret $L(\boldsymbol{\theta}|\hat{\boldsymbol{\theta}})$ as the probability distribution of $\boldsymbol{\theta}$. We know nothing of either $P(\boldsymbol{\theta})$ or $P(\hat{\boldsymbol{\theta}})$, so in the best Bayesian tradition assume that both are constants, in which case $L(\boldsymbol{\theta}|\hat{\boldsymbol{\theta}})$ is equal to $L(\hat{\boldsymbol{\theta}}|\boldsymbol{\theta})$, as both are normalized. The results will be the same as those given by the approach of this section.

‡Or can be linearized by expanding the model in a Taylor series about the best-fit values of the parameters, $\hat{\boldsymbol{\theta}}$. This will only work if the uncertainties in the parameters are small enough that a linear expansion is valid; see problem 71.

to calculate $L(\boldsymbol{\theta}|\hat{\boldsymbol{\theta}})$, as we shall see when we have covered the appropriate theory in section 11.5. If the model is simple enough, and the errors are sufficiently well behaved, it may be possible to obtain similar analytical results in other cases, but usually we are forced to fall back upon Monte Carlo methods.

We have only one set of data, but we can use a computer to generate more. We have assumed that the true parameters of our system are $\boldsymbol{\Theta}$, so if we understand our errors it's easy enough to write a program to create fake datasets. We can repeat our original analysis and estimate θ for each dataset; let us call these estimates $\hat{\boldsymbol{\theta}}'$. Once we have our set of values of $\hat{\theta}'$ we can find their distribution $L(\hat{\boldsymbol{\theta}}'|\boldsymbol{\Theta})$. Because $\hat{\theta}'$ is an ML estimate the mode of this distribution will be at $\boldsymbol{\Theta}$, but as ML estimators can be biased there is no guarantee that its mean will also be $\boldsymbol{\Theta}$. If you care about this bias you might want to reconsider the original choice $\boldsymbol{\Theta} = \hat{\theta}$ and iterate until you find a $\boldsymbol{\Theta}$ satisfying $\langle\boldsymbol{\Theta}\rangle = \hat{\theta}$, but this will no longer be an ML estimator.

We can now find a number of surfaces $L(\hat{\boldsymbol{\theta}}'|\boldsymbol{\Theta}) = c_\alpha$, adjusting the value of c_α until we find a surface that encloses the desired fraction of the fake points; this defines our confidence region $\mathcal{R}$. Note that the surfaces will not in general be elliptical; elliptical confidence regions are common only because they are produced by the linear-and-Gaussian theory that we were unable to apply to our problem.

It is clear that the parameters estimated from our fake datasets can be used to find confidence regions for a subset of the parameters by simply ignoring the uninteresting parameters while obtaining the confidence surfaces, i.e., by working in a lower dimensional subspace. We can still use a surface of constant $L(\boldsymbol{\theta}|\hat{\boldsymbol{\theta}})$ to define our confidence region.

As an alternative to this procedure we can use a bootstrap (section 6.5); not only are bootstraps applicable even if we don't know the distribution of the errors, but they also allow us to investigate two distinct sources of error: those due to errors in the measurements, and those due to our choice of sample points.

To investigate the influence of measurement errors we can form a bootstrap estimate of ϵ's distribution by subtracting our best model from the data, resulting in n values sampled from ϵ; standard bootstrap procedures then lead to an estimate of the uncertainties in $\hat{\boldsymbol{\theta}}$. If there's reason to believe that the errors aren't the same for each data point, this simple approach must be modified — for example, if you thought that the variance was proportional to the observation, then the bootstrap values of ϵ could be suitably scaled.

An alternative to this bootstrap estimate of the errors is to simply apply bootstrap resampling to the data points themselves, allowing us to investigate not only the importance of random errors, but also the very choice of our sample. For example, in problem 61 we measured a velocity for every star that was bright enough, and we would like to know how much the vagaries of stellar evolution, and thus our choice of stars, affected our conclusions.†

References

1: *K&S* 17.14
2: *K&S* 17.15
3: *K&S* 17.17
4: *K&S* Ex. 17.6
5: *K&S* 18.9, 18.10, 18.15
6: *LGG*
7: *K&S* 19.2
8: *Bevington* 5.1
9: *Bevington II* 6.2
10: *NR* 14.5

† I must confess that we did not, in fact, carry out any bootstrap calculations.

11. Least Squares Fitting for Linear Models

11.1. Relationship to MLE

We have seen that MLE reduces to minimizing a χ^2 variable in the case where the errors are known and follow a Gaussian distribution. The function to be minimized is

$$S = \sum_{ij} (y_i - f(x_i;\theta)) V_{ij}^{-1} (y_j - f(x_j;\theta))$$

where V is again the *covariance* matrix,

$$V_{ij} \equiv \left\langle (y_i - \langle y_i \rangle)(y_j - \langle y_j \rangle) \right\rangle .$$

The value of θ that minimizes S is written as $\hat{\theta}$ and is referred to as the *least squares estimator* (LSE) of θ.

If all the variances are equal to σ^2, and all the covariances (i.e., the off-diagonal elements) are zero, S reduces to[1]

$$S = \frac{1}{\sigma^2} \sum_i (y_i - f(x_i;\theta))^2 .$$

We are at liberty to use the minimization of S to estimate parameters even when the errors are not Gaussian; in this case LSE and MLE will (in general) give different answers and we must choose between them based on their properties rather than our prejudices.

For linear models LSE has the important property of being unbiased, even for small samples (of course this property carries over to MLE estimation applied to samples with Gaussian errors). In addition, least squares estimators have the lowest variance of any unbiased linear estimator of θ (see problem 65).

We can write a general linear model† as[2]

$$\mathbf{y} = \mathbf{c} + M\boldsymbol{\theta} + \boldsymbol{\epsilon},$$

---

†Linear in the parameters, that is, so a quadratic function such as $y_i = a + bx_i^2$ counts as linear. Some functions such as $y_i = ae^{bx_i}$ can be made linear by a suitable change of variables;[2] of course you must worry about the effect of such a transformation upon the distribution of the errors.

where $\mathbf{y}$ is an n-dimensional vector of observations, $\mathbf{c}$ is a vector of known constants, $\boldsymbol{\theta}$ is a k-dimensional vector of parameters to be estimated, M is an $n \times k$ matrix of known coefficients, and $\boldsymbol{\epsilon}$ is an n-dimensional vector of "error" variables with $\langle \boldsymbol{\epsilon} \rangle = \mathbf{0}$ and covariance matrix

$$V(\boldsymbol{\epsilon}) \equiv \langle \boldsymbol{\epsilon}\boldsymbol{\epsilon}^T \rangle$$

($\boldsymbol{\epsilon}^T$ is the transpose of $\boldsymbol{\epsilon}$). I shall in fact omit the $\mathbf{c}$ term in the analysis that follows, which doesn't involve any loss of generality as I can always redefine $\mathbf{y}$ as $\mathbf{y}-\mathbf{c}$. The matrix M is often written as X, as it represents the independent variables in the problem, but I reserve X for other purposes.

11.2. A Simple Example

As an example to illustrate the notation, and because the results are important in their own right, let us consider the problem of fitting a straight line to data where all the errors are in one variable.[3,4,5] We shall return to the case where there are errors in both x and y in section 11.7.

If we wish to fit a straight line our model is $y_i = \theta_1 + \theta_2 x_i + \epsilon_i$, so $\boldsymbol{\theta}$ has two components ($k = 2$), the first column of M is all ones, and the second column consists of the (known) x_i.[6] For convenience let's assume that all the variances are equal to σ^2 and that all of the covariances vanish. Under these conditions (as we shall shortly see) the LSE is

$$\hat{\boldsymbol{\theta}} = (M^T M)^{-1} M^T \mathbf{y},$$

and $\hat{\boldsymbol{\theta}}$'s covariance matrix is

$$W(\hat{\boldsymbol{\theta}}) = \sigma^2 (M^T M)^{-1}.$$

We can write $M = (\mathbf{1}\ \ \mathbf{x})$, so

$$\mathbf{y} = (\mathbf{1}\ \ \mathbf{x}) \begin{pmatrix} \theta_1 \\ \theta_2 \end{pmatrix} + \boldsymbol{\epsilon}$$

and

$$\begin{aligned} \hat{\boldsymbol{\theta}} &= \begin{pmatrix} \mathbf{1}^T\mathbf{1} & \mathbf{1}^T\mathbf{x} \\ \mathbf{x}^T\mathbf{1} & \mathbf{x}^T\mathbf{x} \end{pmatrix}^{-1} \begin{pmatrix} \mathbf{1}^T\mathbf{y} \\ \mathbf{x}^T\mathbf{y} \end{pmatrix} \\ &= \begin{pmatrix} n & \sum x \\ \sum x & \sum x^2 \end{pmatrix}^{-1} \begin{pmatrix} \sum y \\ \sum xy \end{pmatrix} \end{aligned}$$

(all sums run over the index $i = 1, n$)

$$= \frac{1}{n\sum x^2 - (\sum x)^2} \begin{pmatrix} \sum x^2 \sum y - \sum x \sum xy \\ -\sum x \sum y + n \sum xy \end{pmatrix},$$

so that

$$\hat{\theta}_1 = \frac{\sum x^2 \sum y - \sum x \sum xy}{n\sum x^2 - (\sum x)^2}$$

and

$$\begin{aligned} \hat{\theta}_2 &= \frac{n\sum xy - \sum x \sum y}{n\sum x^2 - (\sum x)^2} \\ &= \frac{\sum (x-\bar{x})(y-\bar{y})}{\sum (x-\bar{x})^2} \\ &= \frac{s_{xy}}{s_x^2} \end{aligned}$$

(s_{xy} is the sample covariance), so

$$\hat{\theta}_1 = \bar{y} - \hat{\theta}_2 \bar{x},$$

which is probably a familiar result.

We can also calculate $\hat{\boldsymbol{\theta}}$'s covariance matrix, $W(\hat{\boldsymbol{\theta}})$:[7,8]

$$\begin{aligned} W(\hat{\boldsymbol{\theta}}) &= \sigma^2 (M^T M)^{-1} \\ &= \sigma^2 \left((\mathbf{1}, \mathbf{x})^T (\mathbf{1}, \mathbf{x})\right)^{-1} \\ &= \sigma^2 \begin{pmatrix} n & \sum x \\ \sum x & \sum x^2 \end{pmatrix}^{-1} \\ &= \frac{\sigma^2}{n\sum x^2 - (\sum x)^2} \begin{pmatrix} \sum x^2 & -\sum x \\ -\sum x & n \end{pmatrix} \\ &= \frac{\sigma^2}{\sum (x-\bar{x})^2} \begin{pmatrix} \sum x^2/n & -\bar{x} \\ -\bar{x} & 1 \end{pmatrix}, \end{aligned}$$

so

$$\begin{aligned} V(\hat{\theta}_1) &= \frac{\sigma^2 \sum x^2/n}{\sum (x-\bar{x})^2} \\ V(\hat{\theta}_2) &= \frac{\sigma^2}{\sum (x-\bar{x})^2} \end{aligned}$$

and

$$\operatorname{Cov}(\hat{\theta}_1 \hat{\theta}_2) = \frac{-\bar{x}\sigma^2}{\sum (x-\bar{x})^2}.$$

In a real problem it is not sufficient merely to estimate the $\hat{\theta}_i$ and their covariance matrix; we must also ask whether the fit is good, whether the parameters in the fit are significant, and what confidence intervals we can place upon the $\hat{\theta}_i$. We shall return to these subjects after dealing with some theory.

11.3. Linear Estimators: Theory

We can write S as[9]

$$S = (\mathbf{y} - M\boldsymbol{\theta})^T V^{-1} (\mathbf{y} - M\boldsymbol{\theta}),$$

which is minimized (differentiating with respect to "$\boldsymbol{\theta}$", i.e., each of the θ_i in turn) when $2M^T V^{-1}(\mathbf{y} - M\boldsymbol{\theta}) = 0$, i.e., the LSE is given by

$$\hat{\boldsymbol{\theta}} = (M^T V^{-1} M)^{-1} M^T V^{-1} \mathbf{y}$$

(providing that $M^T V^{-1} M$ is non-singular).

Problem 63. Prove that $S = (\mathbf{y} - M\boldsymbol{\theta})^T V^{-1}(\mathbf{y} - M\boldsymbol{\theta})$ is minimized by choosing $\hat{\boldsymbol{\theta}} = (M^T V^{-1} M)^{-1} M^T V^{-1} \mathbf{y}$.

Problem 64. Why can't I write $(M^T V^{-1} M)^{-1} = M^{-1} V M^{T^{-1}}$ and deduce that $\hat{\boldsymbol{\theta}} = M^{-1}\mathbf{y}$?

LSEs of linear functions of $\boldsymbol{\theta}$ are unbiased:[10]

$$\begin{aligned}\hat{\boldsymbol{\theta}} &= (M^T V^{-1} M)^{-1} M^T V^{-1} (M\boldsymbol{\theta} + \boldsymbol{\epsilon}) \\ &= \boldsymbol{\theta} + (M^T V^{-1} M)^{-1} M^T V^{-1} \boldsymbol{\epsilon}.\end{aligned}$$

V and M are known matrices and $\langle \boldsymbol{\epsilon} \rangle = \mathbf{0}$, so $\langle \hat{\boldsymbol{\theta}} \rangle = \boldsymbol{\theta}$. Note that in deriving this we have made no assumptions about $V(\boldsymbol{\epsilon})$.

We can also calculate $\hat{\boldsymbol{\theta}}$'s covariance:[11]

$$\begin{aligned}W(\hat{\boldsymbol{\theta}}) &= \left\langle (\hat{\boldsymbol{\theta}} - \boldsymbol{\theta})(\hat{\boldsymbol{\theta}} - \boldsymbol{\theta})^T \right\rangle \\ &= \langle (M^T V^{-1} M)^{-1} M^T V^{-1} \boldsymbol{\epsilon}\boldsymbol{\epsilon}^T V^{-1} M (M^T V^{-1} M)^{-1} \rangle \\ &= (M^T V^{-1} M)^{-1} M^T V^{-1} \langle \boldsymbol{\epsilon}\boldsymbol{\epsilon}^T \rangle V^{-1} M (M^T V^{-1} M)^{-1} \\ &= (M^T V(\boldsymbol{\epsilon})^{-1} M)^{-1}.\end{aligned}$$

If $V = \sigma^2 \mathbf{1}$ this reduces to $\sigma^2 (M^T M)^{-1}$. The next problem shows that LSEs have the lowest variance of any unbiased estimator that is linear in $\mathbf{y}$.

Problem 65. Consider some linear function of $\mathbf{y}$, $\mathbf{t} = T\mathbf{y}$. If $\mathbf{t}$ is an unbiased estimator for $\boldsymbol{\theta}$, show that $TM = I$. Calculate $\mathbf{t}$'s variance, and show that it cannot be less than the variance of $\hat{\boldsymbol{\theta}}$. (*Hint:* write $\mathbf{t} = \hat{\boldsymbol{\theta}} + (\mathbf{t} - \hat{\boldsymbol{\theta}})$.)[12]

11.4. Goodness of Fit

We have derived the covariance of $\hat{\boldsymbol{\theta}}$ on the assumption that we know V, the covariance matrix of the errors. We are also free to estimate V from the data, and a comparison of the two provides a measure of how good the model is. If we don't know V we can turn this argument around and, by asserting that the model is good, estimate V.

Let us adopt the null hypothesis "The model is perfect," in which case the residuals $\boldsymbol{\delta} \equiv \mathbf{y} - M\hat{\boldsymbol{\theta}}$ are only non-zero due to the measurement errors $\boldsymbol{\epsilon}$. We can use a statistical test to see if the observed $\boldsymbol{\delta}$ are consistent with the known $\boldsymbol{\epsilon}$; if they are too large we must reject H_0. Of course we need to know the distribution of the $\boldsymbol{\epsilon}$ in order to make any progress. We shall assume that they are multivariate Gaussian with zero mean and covariance matrix V; if anyone dares challenge us we shall appeal to the central limit theorem, and, indeed, if many independent terms contribute to the errors our position might even be defensible.

The residuals are given by[13]

$$\begin{aligned}\boldsymbol{\delta} &\equiv \mathbf{y} - M\hat{\theta} \\ &= (I_n - M(M^TV^{-1}M)^{-1}M^TV^{-1})\mathbf{y}\end{aligned}$$

(I_n is an identity matrix of rank n). Substituting $\mathbf{y} = M\boldsymbol{\theta} + \boldsymbol{\epsilon}$ gives

$$\boldsymbol{\delta} = (I_n - M(M^TV^{-1}M)^{-1}M^TV^{-1})\boldsymbol{\epsilon},$$

so the $\boldsymbol{\delta}$ are a linear function of the $\boldsymbol{\epsilon}$, and therefore they too follow a multivariate Gaussian distribution.

There are n data points and k parameters estimated from the data, so if we calculate

$$X^2 \equiv \boldsymbol{\delta}^TV^{-1}\boldsymbol{\delta} = (\mathbf{y} - M\hat{\boldsymbol{\theta}})^TV^{-1}(\mathbf{y} - M\hat{\boldsymbol{\theta}})$$

it's the sum of squares of $n - k$ independent, normalized, Gaussian variables; in other words, X^2 is a χ^2_{n-k} variable (see the next problem). The variable that I have called X^2 is often called χ^2, a usage that has nothing to recommend it and which can lead to great confusion.

Problem 66. Prove that X^2 is a χ^2_{n-k} variable. (*Hint:* see section 4.2 and problem 22.)

We can now use the observed value of X^2 to perform the test suggested at the top of this section: calculate X^2 and see if it is larger than would be expected for a χ^2_{n-k} variable; if it is (at a stated significance level) we must reject the model.[14,15] You can either perform the test mentally by noting that X^2 should not exceed $n-k$ by more than a few $\sqrt{2(n-k)}$ (problem 21), or by using tables of χ^2 to see exactly at what confidence level we can reject the model. Of course, if we can only reject the model at the 50% level we will accept it instead.

Problem 67. I find a value of $X^2 = 47.1$ when fitting a model with 3 parameters to 30 observations. At what confidence level can I reject the model?

If you don't like matrix notation, and for the case where the errors are independent, the formula for X^2 reduces to

$$X^2 = \sum \frac{(y_i - (M\hat{\theta})_i)^2}{\sigma_i^2},$$

where $(M\hat{\theta})_i$ is the model value for the ith point.

We can use this result to estimate σ^2 if we are prepared to *assume* that the model fit is good:[16] if we set

$$s^2 = \frac{1}{n-k} \sum (y_i - (M\hat{\theta})_i)^2,$$

then taking expectation values on each side shows that s^2 is unbiased for σ^2. If we are only estimating the mean this reduces to

$$s^2 = \frac{1}{n-1} \sum (y_i - (M\hat{\theta})_i)^2,$$

as you should expect.

Problem 68. Show that $ns_y^2(1 - r_{xy}^2)/(n-2)$ is an unbiased estimator of σ^2 for the model of section 11.2.

Problem 69. If $V = \sigma^2 I$, prove that $\langle \delta(\hat{\theta} - \theta)^T \rangle = 0$, i.e., that the residuals and the uncertainties in θ are uncorrelated.

11.5. Errors in the Parameters

We are now in a position to estimate the errors in our models. The theory that we develop will only apply to models that are indeed linear and for which the errors are indeed Gaussian. Furthermore, the fit had better be acceptable or the results that we develop will be meaningless.

If we want to estimate the uncertainties in the model parameters $\hat{\boldsymbol{\theta}}$, we are interested in the difference between our estimates and the true values $\boldsymbol{\theta}$. We know that

$$\hat{\boldsymbol{\theta}} - \boldsymbol{\theta} = (M^T V^{-1} M)^{-1} M^T V^{-1} \boldsymbol{\epsilon},$$

so $(\hat{\boldsymbol{\theta}} - \boldsymbol{\theta})_i$ is a linear combination of Gaussian variables and thus follows a multivariate Gaussian distribution. We can calculate the expectation value and covariance matrix:[17]

$$\langle \hat{\boldsymbol{\theta}} - \boldsymbol{\theta} \rangle = (M^T V^{-1} M)^{-1} M^T V^{-1} \langle \boldsymbol{\epsilon} \rangle = 0$$

(i.e., $\hat{\boldsymbol{\theta}}$ is an unbiased estimator of $\boldsymbol{\theta}$) and

$$\begin{aligned} \left\langle (\hat{\boldsymbol{\theta}} - \boldsymbol{\theta})(\hat{\boldsymbol{\theta}} - \boldsymbol{\theta})^T \right\rangle &\equiv W \\ &= (M^T V^{-1} M)^{-1}, \end{aligned}$$

where we have reintroduced W as the covariance matrix for the k elements of θ. The probability distribution of $\hat{\boldsymbol{\theta}} - \boldsymbol{\theta}$ is thus

$$P(\hat{\boldsymbol{\theta}} - \boldsymbol{\theta}) = \frac{1}{(2\pi)^{k/2} |W|^{1/2}} e^{-(\hat{\boldsymbol{\theta}} - \boldsymbol{\theta})^T W^{-1} (\hat{\boldsymbol{\theta}} - \boldsymbol{\theta})/2}.$$

Note that in general W will not be diagonal and the errors in the estimated parameters will be correlated. If we imagine plotting surfaces of constant probability in k-dimensional parameter space they will be similar ellipsoids centred on $\hat{\boldsymbol{\theta}}$.[18]

As an alternative to evaluating W directly, the covariance matrix can be expressed in terms of X^2: consider

$$X^2 = (\mathbf{y} - M\hat{\boldsymbol{\theta}})^T V^{-1} (\mathbf{y} - M\hat{\boldsymbol{\theta}})$$

or, switching to using the summation convention,

$$= (y_i - M_{ij}\hat{\theta}_j) V^{-1}_{ik} (y_k - M_{kl}\hat{\theta}_l),$$

so

$$\frac{\partial X^2}{\partial \hat{\theta}_\alpha} = -2M_{i\alpha}V_{ik}^{-1}(y_k - M_{kl}\theta_l)$$

and

$$\frac{1}{2}\frac{\partial^2 X^2}{\partial \hat{\theta}_\alpha \partial \hat{\theta}_\beta} = M_{i\alpha}V_{ik}^{-1}M_{k\beta}$$

$$= (M^T V^{-1} M)_{\alpha\beta},$$

so we can also express the covariance matrix as

$$W = \left(\frac{1}{2}\frac{\partial^2 X^2}{\partial \hat{\boldsymbol{\theta}} \partial \hat{\boldsymbol{\theta}}}\right)^{-1}.$$

For our linear model,

$$\frac{1}{2}\frac{\partial^2 X^2}{\partial \boldsymbol{\theta} \partial \boldsymbol{\theta}} = -\frac{\partial^2 \ln L}{\partial \boldsymbol{\theta} \partial \boldsymbol{\theta}},$$

so this is a natural generalization of the MVB of section 10.1.

For non-linear models we are forced to use some numerical minimization scheme to find $\hat{\boldsymbol{\theta}}$ (we have presumably now expanded about $\hat{\boldsymbol{\theta}}$ in order to estimate our errors, having confirmed that such an expansion is justified). A fortunate byproduct of such schemes is that they generally require us to calculate or estimate the second derivatives of X^2 with respect to $\boldsymbol{\theta}$, in which case we have already completed the work involved in finding the covariance matrix.

The quantity

$$X_k^2 = (\hat{\boldsymbol{\theta}} - \boldsymbol{\theta})^T W^{-1} (\hat{\boldsymbol{\theta}} - \boldsymbol{\theta})$$

has k degrees of freedom and therefore follows a χ_k^2 distribution (see problem 66 for an example of the sort of calculation required to prove this). This χ_k^2 distribution should not be confused with the χ_{n-k}^2 distribution resulting from fitting the data to the model.

You can also see this by considering the Taylor expansion of X^2 about its value when k parameters have been estimated from the data. We have already calculated the required derivatives:

$$X^2 = X^2\Big|_{\boldsymbol{\theta}=\hat{\boldsymbol{\theta}}} + (\hat{\boldsymbol{\theta}} - \boldsymbol{\theta})^T W^{-1} (\hat{\boldsymbol{\theta}} - \boldsymbol{\theta})$$

(all higher terms vanish). If there are n data points, the left-hand side has n degrees of freedom, as it is evaluated for a given set of parameters

θ, while $X^2|_{\theta=\hat\theta}$ has only $n-k$, as it is the value of X^2 obtained while estimating $\hat\theta$. X_k^2 must therefore follow a $\chi^2_{n-(n-k)} = \chi_k^2$ distribution, as claimed. (To complete the proof we'd have to prove that the two terms on the right-hand side of the equation are independent; the calculation is similar to that of problem 66 or 78.)

If we want to choose a confidence region $\mathcal{R}$ for the parameters $\hat\theta$ we should obviously choose one bounded by a surface of constant probability, i.e., a surface of constant X^2 (and thus of constant $\Delta X^2 \equiv X^2 - X^2|_{\theta=\hat\theta} = X_k^2$). Because we know that X_k^2 is a χ_k^2 variable we can easily choose the appropriate constant using tables of χ^2. For example, if we have only one parameter ($k = 1$) we know that we will obtain a value of X_k^2 as large as $1.96^2 = 3.84$ only 5% of the time (remember that a χ_1^2 variate is just the square of a Gaussian). Further values are given in the accompanying table, in which the strange values of $1-\alpha$ (68.3%, 95.4%, 99.73%) are chosen to correspond to the probabilities of being one, two, or three standard deviations away from the mean in the familiar Gaussian case ($k = 1$).

	k					
$1-\alpha$	1	2	3	4	5	6
68.3%	1.00	2.30	3.53	4.72	5.89	7.04
90%	2.71	4.61	6.25	7.78	9.24	10.6
95.4%	4.00	6.17	8.02	9.70	11.3	12.8
99%	6.63	9.21	11.3	13.3	15.1	16.8
99.73%	9.00	11.8	14.2	16.3	18.2	20.1
99.99%	15.1	18.4	21.1	23.5	25.7	27.8

X_k^2 as a function of confidence level $1-\alpha$ and number of degrees of freedom k.

As an example, consider the case $k = 2$ and ask for the 99% confidence region. We can plot θ_1 against θ_2 and look for a region $\mathcal{R}$. We know that

$$\begin{aligned} X_2^2 &= (\boldsymbol\theta - \hat{\boldsymbol\theta})^T W^{-1}(\boldsymbol\theta - \hat{\boldsymbol\theta}) \\ &= W_{11}^{-1}(\theta_1-\hat\theta_1)^2 + 2W_{12}^{-1}(\theta_1-\hat\theta_1)(\theta_2-\hat\theta_2) + W_{22}^{-1}(\theta_2-\hat\theta_2)^2 \end{aligned}$$

is a χ_2^2 variable, so (referring to the table) X_2^2 would be expected to be less than 9.21, 99% of the time. Our confidence region is bounded by the ellipse $X_2^2 = 9.21$. What if we want to express our region not as an ellipse, but as separate constraints on $\hat{\theta}_1$ and $\hat{\theta}_2$? We simply construct the rectangle that just bounds our ellipse, and use its sides to define limits on $\hat{\boldsymbol{\theta}}$. Note that this is a larger region than the ellipse.

If we are only interested in some subset of the parameters we need to integrate $\Delta\boldsymbol{\theta}$'s probability distribution over all possible values of the uninteresting ones' parameters, leaving behind the p.d.f. of the others. Because of the nature of Gaussian integrals, it is quite clear that the resulting p.d.f. is also multivariate Gaussian and completely characterized by its covariance matrix (as all of the means are zero). A moment's reflection will reveal that we already know the elements of the covariance matrix, as they are simply the variances and covariances of the parameters in question, which we can read from the corresponding elements of W. To find a confidence interval for r parameters, the procedure is first to find W (possibly by inverting the matrix $(1/2)\partial^2 X^2/\partial\boldsymbol{\theta}\partial\boldsymbol{\theta}$), and then to construct the covariance matrix W_r for the parameters of interest by copying the appropriate entries in W. If I then calculate

$$X_r^2 = (\hat{\boldsymbol{\theta}}_r - \boldsymbol{\theta}_r)^T W_r^{-1} (\hat{\boldsymbol{\theta}}_r - \boldsymbol{\theta}_r)$$

it will be distributed as χ_r^2, and I can proceed to set limits on the r parameters just as before. In the 2-parameter example of the previous paragraph, if I want to derive a confidence interval for $\hat{\theta}_1$ I extract the appropriate element from W, namely σ_{11}^2, and calculate X_1^2, which is now a χ_1^2 variable:

$$X_1^2 = \frac{(\theta_1 - \hat{\theta}_1)^2}{\sigma_{11}^2}.$$

My table tells me that the 99% point for one degree of freedom is 6.63, so the confidence region is $X_1^2 < 6.63$, i.e.,

$$|\theta_1 - \hat{\theta}_1| < \sqrt{6.63}\sigma_{11}$$

or

$$\hat{\theta}_1 - 2.57\sigma_{11} < \theta_1 < \hat{\theta}_1 + 2.57\sigma_{11}.$$

This is just the value that you would derive from tables of the Gaussian distribution (as a χ_1^2 variate is just the square of a Gaussian).

Problem 70. Consider a model with two parameters θ_1 and θ_2 with $\hat{\boldsymbol{\theta}} = \mathbf{0}$ and a covariance matrix

$$W = \begin{pmatrix} a & b \\ b & c \end{pmatrix}^{-1}.$$

Write down the p.d.f. for $\boldsymbol{\theta}$, and integrate over all θ_2 to show that the distribution of θ_1 is a Gaussian with mean zero and variance $(a - b^2/c)^{-1}$. Confirm this result by applying the theory of the last paragraph.

Problem 71. I have taken a picture of a certain star, and now I want to measure its position and brightness. If the intensity I of points in the image is measured in numbers of photons, what is the variance of I? (*Hint:* see problem 8.) I know that at a distance of r from the centre $\mathbf{X} \equiv (X, Y)$ the intensity is

$$I = S + I_c f(r; \mathbf{X}),$$

where I_c is the star's (unknown) central intensity, f is a known function describing the shape of the stellar image, and S is the (unknown) brightness of the night sky. If I have measurements of the intensity I_i at n points, write down the log-likelihood that I could use to make an ML estimate of the four unknowns $(S, I_c, X, Y) \equiv \boldsymbol{\theta}$; let us write the ML estimate as $(S_0, I_{c0}, X_0, Y_0) \equiv \boldsymbol{\theta}_0$.

In order to estimate the errors I decide to linearize the model about $\boldsymbol{\theta}_0$; rewrite the model in the form

$$\mathbf{y} = M(\boldsymbol{\theta} - \boldsymbol{\theta}_0) + \boldsymbol{\epsilon} + O\left((\boldsymbol{\theta} - \boldsymbol{\theta}_0)^2\right),$$

neglect the quadratic term, and hence find $\boldsymbol{\theta}$'s covariance matrix (assume that $S \gg 1$ and that the star occupies only a small part of the image).

If the star is faint ($I_c \ll S$) and the profile f is Gaussian, what does the covariance matrix reduce to? What is the variance of the total number of photons detected? In the same limit give an approximate condition under which the quadratic terms can really be ignored. (*Hint:* approximate all sums as integrals.)

11.6. Fitting Models to Data with Non-Gaussian Errors

Sometimes the error distribution will not be Gaussian, or at least you don't want to assume that it is.[19] If large errors are not especially unlikely, a fit based on a Gaussian will be far too much influenced by outlying points, as they would never (well, hardly ever) have arisen by chance from a Gaussian error distribution. To proceed, guess a distribution for the errors, e.g.,

$$dF = \frac{1}{2a} e^{-|x|/a}$$

in which case the ML estimator is given by minimizing[20]

$$\sum_i |\mathbf{y} - M\boldsymbol{\theta}|_i$$

rather than

$$\sum_i (\mathbf{y} - M\boldsymbol{\theta})_i^2.$$

In fact, we are free to minimize *any* function of the residuals, and the only question that arises is "is it efficient?". Usually there is no good answer, as the true distribution of the errors is unknown; we use something like "minimize the absolute deviations" more because we have a feeling that it is a sensible thing to do than from any sure statistical footing.

11.7. Fitting Models with Errors in Both x and y

If we have errors in the measurements of both x and y we can't simply use our previous results.[21,22] Specifically, if

$$x_i = \xi_i + \delta_i$$
$$y_i = \eta_i + \epsilon_i$$

with

$$\langle \delta \rangle = \langle \epsilon \rangle = \langle \delta\epsilon \rangle = 0$$
$$\langle \delta^2 \rangle = \phi_x$$
$$\langle \epsilon^2 \rangle = \phi_y,$$

and if our model, which is of course expressed in the (unknown) ξ_i and η_i, is

$$\eta_i = a\xi_i + b,$$

then

$$y_i = ax_i + b - (a\delta - \epsilon),$$

which is not a simple regression problem, as the error term is correlated with x:[23]

$$\langle x(a\delta - \epsilon)\rangle = \langle(\xi + \delta)(a\delta - \epsilon)\rangle = a\langle\delta^2\rangle = a\phi_x \neq 0.$$

Problem 72. If δ and ϵ are known to be Gaussian, show that $\epsilon - a\delta$ is an $N(0, \phi_y + a^2\phi_x)$ variable. Hence invent a way to find a confidence region for (a, b).

If we are prepared to assume that the errors δ_i and ϵ_i are Gaussian, we can write down the likelihood of the observation (x_i, y_i):[24]

$$L_i = \frac{1}{2\pi\sqrt{\phi_x\phi_y}} e^{-(x_i-\xi_i)^2/2\phi_x-(y_i-a\xi_i-b)^2/2\phi_y},$$

so

$$\ln L = -n\ln(2\pi) - \frac{n}{2}\ln\left(\phi_x\phi_y\right) - \frac{1}{2\phi_x}\sum(x_i-\xi_i)^2 - \frac{1}{2\phi_y}\sum(y_i-a\xi_i-b)^2.$$

We can now differentiate the likelihood with respect to the parameters in the model a, b, the n ξ_i, ϕ_x, and ϕ_y; setting the derivatives to zero gives us

$$\sum\hat{\xi}_i(y_i - \hat{a}\hat{\xi}_i - \hat{b}) = 0, \quad (1)$$

$$\sum(y_i - \hat{a}\hat{\xi}_i - \hat{b}) = 0, \quad (2)$$

$$\hat{\phi}_y(x_i - \hat{\xi}_i) + \hat{a}\hat{\phi}_x(y_i - \hat{a}\hat{\xi}_i - \hat{b}) = 0, \quad (3)$$

$$n\hat{\phi}_x - \sum(x_i - \hat{\xi}_i)^2 = 0, \quad (4)$$

and

$$n\hat{\phi}_y - \sum(y_i - \hat{a}\hat{\xi}_i - \hat{b})^2 = 0. \quad (5)$$

Equations 3 and 4 show that

$$\hat{\phi}_y^2 = \frac{\hat{a}^2 \hat{\phi}_x}{n} \sum (y_i - \hat{a}\hat{\xi}_i - \hat{b})^2,$$

and a glance at equation 5 then shows that

$$\hat{\phi}_y = \hat{a}^2 \hat{\phi}_x.$$

Let us write $\lambda \equiv \hat{a}^2$ in this relation, for reasons that will shortly become clear, and return our attention to equation 3. After a little rearrangement this becomes

$$\hat{\xi}_i = \frac{1}{\lambda + \hat{a}^2} \left(\lambda x_i + \hat{a}(y_i - \hat{b}) \right)$$

and therefore

$$\begin{aligned} y_i - \hat{a}\hat{\xi}_i - \hat{b} &= \frac{\lambda}{\lambda + \hat{a}^2} (y_i - a x_i - \hat{b}) \\ &= \frac{\lambda}{\lambda + \hat{a}^2} (y_i - \bar{y} - a(x_i - \bar{x})), \end{aligned}$$

so (using equation 2)

$$\bar{y} = \hat{a}\bar{x} + \hat{b}.$$

Substituting the formula for $\hat{\xi}_i$ into equation 1 (and fiddling for a few lines of algebra) then gives

$$\hat{a}^2 s_{xy} + \hat{a}(\lambda s_x^2 - s_y^2) - \lambda s_{xy} = 0 \qquad (6),$$

where s_{xy} is x and y's covariance, and s_x^2 and s_y^2 are their variances (I realize that as $\lambda \equiv \hat{a}^2$ this equation reduces to $\hat{a} = s_y / s_x$, but be patient — we will need the full quadratic in a moment). When you solve the quadratic for $\hat{a}$ you should take the square root to be positive.

Problem 73. Why should you take the square root to be positive?

With most of the algebra behind us, let us substitute $\lambda = \hat{a}^2$ and see what we have learned. It proves helpful to consider the x_i's, y_i's, and ξ_i's as lying in an n-dimensional space and to write them as vectors; the equation for $\hat{\xi}_i$ then becomes[25]

$$\hat{\boldsymbol{\xi}} = \frac{1}{2} \left(\mathbf{x} + \frac{1}{\hat{a}} \left(\mathbf{y} - \hat{b} \right) \right).$$

If I maximize the likelihood over all parameters except $\boldsymbol{\xi}$, the log-likelihood at the point $\boldsymbol{\xi}$ (not necessarily $\hat{\boldsymbol{\xi}}$) is

$$\ln L = C - n \ln \left(|\mathbf{x} - \boldsymbol{\xi}| \cdot |\boldsymbol{\xi} - (\mathbf{y} - \hat{b})/\hat{a}| \right),$$

where $| \cdots |$ is, as usual, the length of a vector.

Now consider a straight line joining $\mathbf{x}$ to $(\mathbf{y} - \hat{b})/\hat{a}$; for any point on this line

$$|\mathbf{x} - \boldsymbol{\xi}| + |\boldsymbol{\xi} - (\mathbf{y} - \hat{b})/\hat{a}| = |\mathbf{x} - (\mathbf{y} - \hat{b})/\hat{a}| = \text{constant},$$

so the log-likelihood is *minimized* at the midpoint, which is just $\hat{\boldsymbol{\xi}}$ — our MLE of the model parameters seems to be not a maximum but a minimum likelihood estimator.

To see that this isn't true, consider a point lying in the hyperplane that's perpendicular to the line joining $\mathbf{x}$ to $(\mathbf{y} - \hat{b})/\hat{a}$ and that passes through $\hat{\boldsymbol{\xi}}$. For such a point $|\mathbf{x} - \boldsymbol{\xi}| > |\mathbf{x} - \hat{\boldsymbol{\xi}}|$ and $|\boldsymbol{\xi} - (\mathbf{y} - \hat{b})/\hat{a}| > |\hat{\boldsymbol{\xi}} - (\mathbf{y} - \hat{b})/\hat{a}|$, and the likelihood is smaller still; the point $\hat{\boldsymbol{\xi}}$ is a saddle-point of the likelihood.

It isn't hard to see what has gone wrong. If we write

$$S_x^2 \equiv \frac{1}{n} \sum (x_i - \xi_i)^2$$

and

$$S_y^2 \equiv \frac{1}{n} \sum (y_i - a\xi_i - b)^2,$$

and drop the constant term $-n \ln(2\pi)$, the likelihood becomes

$$-\frac{2}{n} \ln L = \ln \phi_x + \ln \phi_y + \frac{S_x^2}{\phi_x} + \frac{S_y^2}{\phi_y}.$$

Now let $\phi_x \to 0$, so $x_i \to \xi_i$ and $S_x \to 0$. The value of $\ln L$ depends on exactly how the limit is taken; if $S_x^2 = -\phi_x(2k/n + \ln \phi_x)$ the limit is k, which we may choose to be $-\infty$, $+\infty$, or any value in between, irrespective of the value of a and b.[26] Clearly there is no ML estimator for this model.

The problem is that we can adjust the ξ_i to minimize S_x^2 without worrying about the effects on S_y^2.[27] In an attempt to find a satisfactory

MLE for the model parameters, let us tie S_x and S_y together by assuming a value for λ rather than by trying to estimate it from the data. The likelihood then becomes (dropping a term in $\ln\lambda$)

$$-\frac{2}{n}\ln L = 2\ln\phi_x + \frac{1}{\phi_x}\left(S_x^2 + \frac{S_y^2}{\lambda}\right).$$

As $\phi_x \to 0$ the term $S_x^2 + S_y^2/\lambda$ remains finite and $\ln L \to -\infty$; we have removed the troubling behaviour at $\phi_x = 0$.

As we are now fixing the ratio of ϕ_y/ϕ_x we mustn't estimate both variances from the data; if we choose to forego estimating ϕ_y, then equation 5 above disappears and equation 4 changes, but you will find that as we only used them when finding that $\phi_y = \hat{a}^2\phi_x$, and as we cunningly wrote $\lambda \equiv \hat{a}^2$ in this equation, all of our previous results for the parameters of the model hold. In equation 6 note that as $\phi_x \to 0$, $\lambda \to \infty$ and we recover the usual result that $\hat{a} = s_{xy}/s_x^2$.

As we know the value of λ we may as well take it to be unity (i.e., we could scale the data), in which case the likelihood becomes

$$L \propto \phi^{-n} e^{-\sum(\delta^2+\epsilon^2)/2\phi},$$

so the ML estimates of $\hat{a}$ and $\hat{b}$ are achieved by minimizing the perpendicular distances from the data points to the fitted line.

Let us estimate ϕ_x. Equations 4 and 5 are replaced by the single equation

$$\phi_x = \frac{1}{2n}\left(\sum(x_i - \hat{\xi}_i)^2 + \frac{1}{\lambda}\sum(y_i - \hat{a}\hat{\xi}_i - \hat{b})^2\right).$$

Knowing $\hat{\xi}_i$, and using equation 3 again to relate $x_i - \hat{\xi}_i$ to $y_i - \hat{a}\hat{\xi}_i - \hat{b}$, we can simplify this to

$$\hat{\phi}_x = \frac{1}{2\,(\lambda + \hat{a}^2)}\left(\hat{a}^2 s_x^2 - 2\hat{a}s_{xy} + s_y^2\right).$$

It's simple enough to show that $\hat{a}$ and $\hat{b}$ are consistent estimators,[28] but our estimator for ϕ_x may readily be shown to be inconsistent; in fact,

$$\lim_{n\to\infty}\hat{\phi}_x = \frac{1}{2}\phi_x.$$

How does this come about? After all, didn't I claim that ML estimators are consistent? The problem is that we are not really estimating ϕ_x from as large a sample as we thought, as we are also estimating the ξ_i from the pairs of observations (x_i, y_i). The number of degrees of freedom is not $2n$ but $2n - n - 2$, and we indeed find that

$$\frac{2n}{n-2}\hat{\phi}_x$$

is consistent for ϕ_x.

Problem 74. Show that our estimator $\hat{a}$ is consistent for a, but that $\lim_{n\to\infty} \hat{\phi}_x = \frac{1}{2}\phi_x$. (*Hint:* do not be afraid to introduce ξ's unknown variance s_ξ^2.)

References

1: *K&S* 19.3
2: *K&S* 19.4
3: *K&S* Ex. 19.3
4: *Bevington* 6.2
5: *Bevington II* 6.2
6: *Sachs* 5.42
7: *K&S* 19.5
8: *Sachs* 5.4.3
9: *K&S* 19.4, 19.17
10: *K&S* 19.5
11: *K&S* 19.5
12: *K&S* 19.6
13: *K&S* 19.9
14: *Bevington* 10.1
15: *Bevington II* 11.1
16: *K&S* 19.9
17: *K&S* 19.5
18: *NR* 14.5
19: *NR* 14.6
20: *K&Sv* 28.72
21: *K&S* 29.3
22: *Sachs* 5.1.2
23: *K&S* 29.4
24: *K&S* 29.14
25: *Solari*
26: *K&S* 29.14
27: *K&S* 29.16
28: *K&S* 29.19

12. Hypothesis Testing in the Linear Model

12.1. Significance of Parameters

We have estimated a number of parameters that specify a model, but how many of them are needed? If we have n data points we could fit a model with n free parameters and obtain a perfect fit to the data, but we couldn't realistically hope to convince anyone else that we had done anything useful.

If the errors are Gaussian, and the model linear, we know that we can see how good the fit is by calculating (cf. section 11.4)

$$\begin{aligned} X^2 &= (\mathbf{y} - M\hat{\boldsymbol{\theta}}\boldsymbol{\theta})^T V^{-1}(\mathbf{y} - M\hat{\boldsymbol{\theta}}\boldsymbol{\theta}) \\ &= \sum_i \frac{(y_i - f(x_i|\boldsymbol{\theta}))^2}{\sigma_i^2}, \end{aligned}$$

which follows a χ^2_{n-k} distribution, where k is the number of parameters in the model and we have assumed that the errors are uncorrelated. As we make k larger the value of X^2 will become smaller in a perfectly predictable way until, when $k = n$, X^2 vanishes. This decrease in X^2 is purely the result of removing degrees of freedom, and would occur for any choice of model. We would hope, however, that our model describes the real world so well that X^2 falls *faster* than this — for example, consider the model $y_i = a + bx_i + cx_i^2$ fitted to data that lie very close to a straight line. First let $b = c = 0$, and calculate X^2. We expect it to be large (it's just the variance of y divided by σ^2). Now add the bx_i term, and X^2 will fall dramatically, much more than we would have expected by adding just one degree of freedom to the fit. Adding the cx_i^2 should result in little additional change in X^2. We can exploit this behaviour to ask which (if any) of our model parameters significantly improve the fit.[1]

We have been considering the model

$$\mathbf{y} = M\boldsymbol{\theta} + \boldsymbol{\epsilon},$$

where the details of the model are all contained in the matrix M. Let us consider a modified model

$$\mathbf{y} = MJ\boldsymbol{\theta} + \boldsymbol{\epsilon}$$

and write $N \equiv MJ$ for convenience; let us assume that J has a rank of $k - r$. For example, if I chose to set the first r elements of $\boldsymbol{\theta}$ equal to 0 but otherwise used the same model, J would be a diagonal $k \times k$ matrix with 0 in the first r positions and 1 in the remaining $k - r$. Setting the r known elements of $\boldsymbol{\theta}$ to 0 rather than to some other set of predetermined values involves no loss of generality, as I can always subtract a constant vector from $\mathbf{y}$ before starting the analysis. As J's rank is less than its dimension k we can't simply write the LSE $\hat{\theta}\boldsymbol{\theta}$ as

$$\hat{\theta}\boldsymbol{\theta} = (N^T V^{-1} N)^{-1} N^T V^{-1} \mathbf{y}$$

because $N^T V^{-1} N$ is singular. This is only formally a problem; the singularity comes about because we are not estimating the first r parameters from the data, so let us simply agree to write 0 in $(N^T V^{-1} N)^{-1}$'s first r rows and columns.†If you think about the derivation of the LSE $\hat{\theta}\boldsymbol{\theta}$, you'll see that offending elements resulted from maximizing with respect to the parameters that we have now arbitrarily chosen.

When you solved problem 66 you showed that

$$\begin{aligned} X_M^2 &= \mathbf{x}^T R^T (V^{-1} - V^{-1} M (M^T V^{-1} M)^{-1} M^T V^{-1}) R \mathbf{x} \\ &\equiv \mathbf{x}^T A_M \mathbf{x}, \end{aligned}$$

where the $\mathbf{x}$ are independent $N(0, 1)$ variables, $RR^T = V$, and I have added a subscript M to indicate that we are using the model $\mathbf{y} = M\boldsymbol{\theta}$.

If we write

$$X_N^2 = X_M^2 + \left(X_N^2 - X_M^2\right),$$

the two terms on the right-hand side are independent. This should not be too surprising, as $X_N^2 - X_M^2$ represents the r degrees of freedom that we "gained" by arbitrarily setting r of the $\hat{\theta}\theta_i$ to some favourite value instead of estimating them from the data. Furthermore, both A_M and A_N are idempotent, so X_M^2 and X_N^2 are both χ^2 variables; calculations similar to those of problem 66 reveal that they have $n - k$ and $n - (k - r)$ degrees of freedom respectively.[3]

Problem 75. Show that X_M^2 and $X_M^2 - X_N^2$ are independent.

†This could be achieved in practice by using *singular value decomposition* (SVD) to invert the matrix.[2]

You should note that in this discussion we have been assuming that the difference in the values of X_M^2 and X_N^2 is solely produced by their different number of degrees of freedom rather than due to the peculiar virtue of our model. If our model is in fact a good one the difference will be bigger than expected, so a test based upon the value of $X_N^2 - X_M^2$ will be a test of how justified we were in fitting the extra r parameters from the data.

We have all the ingredients that we need to test hypotheses, providing that we know the covariance matrix V. If we know that it's of the form $V = \sigma^2 I$, which is frequently the case, we can concoct a test without knowing σ: Consider

$$f = \frac{(X_N^2 - X_M^2)/r}{X_M^2/(n-k)},$$

which is of the form

$$\frac{\chi_\alpha^2/\alpha}{\chi_\beta^2/\beta},$$

i.e., f follows an $F_{r,n-k}$-distribution.[4] Our test consists of calculating the value of f and asking for the probability that it would have arisen simply be removing r degrees of freedom from the fit, i.e., the probability that it is drawn from an $F_{r,n-k}$-distribution.[5,6] If the extra r parameters are significant, f will be larger than we would expect based upon counting degrees of freedom, so we should use a 1-sided test (based on the upper tail of the F-distribution) to test the significance of our models; a large value of f allows us to reject the null hypothesis "the added parameter does not significantly improve the fit."

Problem 76. I wish to test the hypothesis, "The population mean is μ" for a sample known to come from a Gaussian population of unknown standard deviation σ. This can be considered a question about whether estimating the mean from the data is significantly better than simply adopting a value of μ, and we can therefore use an F-test in the manner that we have just described. Show that this procedure is equivalent to using a t-test based on the statistic $(\bar{x} - \mu)/(s/\sqrt{n-1})$.

Problem 77. I have a linear model with 10 parameters, fitted to 50 data points whose errors I know to be Gaussian. I fit the model to the data and obtain a value of $X^2 = 95.4$. I then set

5 of my parameters to 0, fit the model to the data, and obtain a value of $X^2 = 127.9$. Am I justified in using all 10 parameters?

We can use a similar procedure to ask whether it's possible to justify adding a further parameter to a model N that already has s parameters. If M has $s+1$ parameters, $\Delta X^2 \equiv X_N^2 - X_M^2$ has only one degree of freedom, so[7]

$$f = (n-s)\frac{\Delta X^2}{X_N^2}$$

follows an $F_{1,n-s}$-distribution (which happens to be the same as a t_{n-s}^2-distribution). We can test whether our extra term significantly improves the fit by using a 1-tailed F-test on the value of f, or equivalently a 2-tailed t-test on the value of $\sqrt{f}$.

If you don't want to know about likelihood ratio tests, you can skip the rest of this section. It derives the same results in a rather different way.[8] For our linear model, the likelihood is given by

$$L = \frac{1}{(2\pi)^{n/2}}\frac{1}{\sigma^n}e^{-(\mathbf{y}-M\boldsymbol{\theta})^2/2\sigma^2},$$

and we know how to estimate $\boldsymbol{\theta}$ and σ:

$$\hat{\boldsymbol{\theta}}\boldsymbol{\theta} = (M^T M)^{-1} M^T \mathbf{y}$$

$$\hat{\boldsymbol{\theta}}\sigma^2 = \frac{1}{n}(\mathbf{y} - M\hat{\boldsymbol{\theta}}\boldsymbol{\theta})^2,$$

so

$$\hat{\boldsymbol{\theta}}L = (2\pi\hat{\boldsymbol{\theta}}\sigma^2 e)^{-n/2}.$$

(Note that these are the ML estimates of $\boldsymbol{\theta}$ and σ^2 and that they are not unbiased; this doesn't matter in this context.)

Now let us adopt $H_0 : \theta_i = \theta_{i,0}$ for $i = 1, \ldots, r$ (so the remaining $k - r$ θ's are to be estimated by ML). If we write $\hat{\hat{\boldsymbol{\theta}}} = (\boldsymbol{\theta}_r, \hat{\boldsymbol{\theta}}_{k-r})$, with the obvious meaning,

$$\hat{\hat{\sigma}}^2 = \frac{1}{n}(\mathbf{y} - M\hat{\hat{\boldsymbol{\theta}}})^2$$

and

$$\hat{\hat{L}} = (2\pi\hat{\hat{\sigma}}^2 e)^{-n/2}.$$

The likelihood ratio is then given by

$$\ell^{2/n} = \frac{\hat{\boldsymbol{\theta}}\sigma^2}{\hat{\hat{\sigma}}^2} \equiv \frac{1}{1+W},$$

where

$$W = \frac{\hat{\hat{\sigma}}^2 - \hat{\theta}\sigma^2}{\hat{\theta}\sigma^2}.$$

It can be shown that if we write $\hat{\hat{\sigma}}^2 = \hat{\theta}\sigma^2 + \tau^2$ then $\hat{\hat{\sigma}}^2$ and τ^2 are independent if H_0 holds (see problem 75). Looking at the definitions given above, it's clear that $\hat{\hat{\sigma}}^2$ is (to within factors of σ^2) a $\chi^2_{n-(k-r)}$ variate, and that $\hat{\theta}\sigma^2$ is χ^2_{n-k}, so τ^2 is χ^2_r (this isn't really all that surprising, as we removed r degrees of freedom from the fit when estimating $\hat{\hat{\sigma}}^2$).

Problem 78. Consider the case where $k = r$, so we may take $\hat{\hat{\theta}} = 0$ without loss of generality. Show that

$$n\hat{\hat{\sigma}}^2 = \mathbf{y}^2 \equiv (\mathbf{y} - M\hat{\theta}\boldsymbol{\theta})^2 + (M\hat{\theta}\boldsymbol{\theta})^2$$

and hence that, in this case, $\hat{\theta}\sigma$ and τ are independent if $\boldsymbol{\theta} = \mathbf{0}$.

Because ℓ is a monotonically decreasing function of W, the LR test is equivalent to rejecting H_0 when W is large. What is the distribution of W? Dividing top and bottom by σ^2 we see that it is the ratio of two independent χ^2 variates with r and $n - k$ degrees of freedom; in other words, it follows an F-distribution. It is therefore easy to carry out the LR test in terms of f, as $(n - k)/r\, W$ is an $F_{r,n-k}$ variate.

12.2. *F*-tests and Linear Regression

In Section 11.2 we derived the least squares solution to the problem of fitting a straight line $y = \theta_1 + \theta_2 x$ to a set of data, namely

$$\hat{\theta}\theta_1 = \bar{y} - \theta_2\bar{x}$$
$$\hat{\theta}\theta_2 = s_{xy}/s_x^2.$$

A convenient and conventional way of seeing how strongly x and y are correlated is to calculate Pearson's correlation coefficient r_{xy}, which always lies in the range $[-1, 1]$.[9,10,11] If the correlation is perfect, $r_{xy} = 1$; if x and y are exactly anticorrelated, $r_{xy} = -1$; and a value of 0 corresponds to a vanishing covariance (and thus, if x and y are Gaussian, to their being independent).

Problem 79. If two variables x and y are independent, and $z = xy$, what is the product-moment (Pearson's) correlation coefficient of x and z?

We can be more precise about the significance of a non-zero value of r_{xy} by using an F-test of the sort that we have just been discussing. Let us calculate

$$
\begin{aligned}
\sigma^2 X_{n-2}^2 &= \frac{1}{n}\sum(y_i - \hat{\theta}\theta_1 - \hat{\theta}\theta_2 x_i)^2 \\
&= \frac{1}{n}\sum((y_i - \bar{y}) - \hat{\theta}\theta_2(x_i - \bar{x}))^2 \\
&= s_y^2 - 2\hat{\theta}\theta_2 s_{xy} + \hat{\theta}\theta_2^2 s_x^2 \\
&= s_y^2 - \frac{s_{xy}}{s_x^2}.
\end{aligned}
$$

It is pretty clear that if I set $\theta_2 = 0$ then I'll find that

$$\hat{\theta}\theta_1 = \bar{y}$$

and

$$
\begin{aligned}
\sigma^2 X_{n-1}^2 &= \frac{1}{n}\sum(y_i - \hat{\theta}\theta_1)^2 \\
&= s_y^2,
\end{aligned}
$$

so

$$
\begin{aligned}
\frac{X_{n-1}^2 - X_{n-2}^2}{X_{n-2}^2} &= \frac{s_{xy}^2}{s_x^2 s_y^2 - s_{xy}^2} \\
&= \frac{r^2}{1 - r^2},
\end{aligned}
$$

where $r^2 \equiv s_{xy}^2 / s_x^2 s_y^2$ is the usual product-moment (Pearson's) correlation coefficient.

The test for H_0 then depends on the fact that[12,13]

$$f = (n-2)\frac{r^2}{1 - r^2}$$

follows an $F_{1,n-2}$-distribution. We are using this to test the hypothesis that $\theta_2 = 0$, and it is clear that large values of f lead to the rejection of H_0, as indeed we showed while discussing the theory of F-tests.

Because an $F_{1,n-2}$-distribution is the same as a t_{n-2}^2 distribution, it may as before be more convenient to carry out the test as a 2-tailed t-test upon the value of $\sqrt{f}$.

Problem 80. Consider the linear model in the case where the errors are Gaussian and $V = \sigma^2 I$. If s^2 is the usual unbiased estimate for σ^2, show that

$$\frac{(\hat{\theta}\theta_i - \theta_i)^2}{s^2(M^{\mathsf{T}}M)^{-1}_{ii}}$$

follows a t^2-distribution with $n - k$ degrees of freedom. Show that this reduces to a known result if I use it to test the hypothesis $H_0 : \theta_2 = 0$ in the model of section 11.2.

12.3. The Distribution of Pearson's Correlation Coefficient

We have just found the distribution of r when x and y are Gaussian and independent (so the population coefficient ρ vanishes). A glance at the definition of the F- or t-distribution shows that r's p.d.f. is

$$dF = \frac{1}{B(\frac{1}{2}, \frac{1}{2}(n-1))}(1 - r^2)^{\frac{1}{2}(n-4)},$$

where $B(x, y)$ is a beta function.[14,15]

When x and y are correlated this simple result doesn't hold. If you calculated r's variance you'd find that (in the large n limit)[16]

$$V(r) = \frac{\rho^2}{n}\left\{\frac{\mu_{22}}{\mu_{11}^2} + \frac{1}{4}\left(\frac{\mu_{40}}{\mu_{20}^2} + \frac{\mu_{04}}{\mu_{02}^2} + \frac{2\mu_{22}}{\mu_{20}\mu_{02}}\right) - \left(\frac{\mu_{31}}{\mu_{11}\mu_{20}} + \frac{\mu_{13}}{\mu_{11}\mu_{02}}\right)\right\},$$

which depends on all the joint moments (of order four or less) of the parent distribution. Rather than use this formula, let us assume that x and y follow a bivariate Gaussian distribution (i.e., a multivariate Gaussian distribution of two variables), in which case we find that[17]

$$V(r) = \frac{1}{n-1}(1 - \rho^2)^2\left(1 + \frac{11\rho^2}{2n}\right) + O(n^{-3}).$$

Under the same assumption it is also possible (if tedious) to show that[18]

$$\langle r \rangle = \rho\left(1 - \frac{1 - \rho^2}{2n} + O(n^{-2})\right),$$

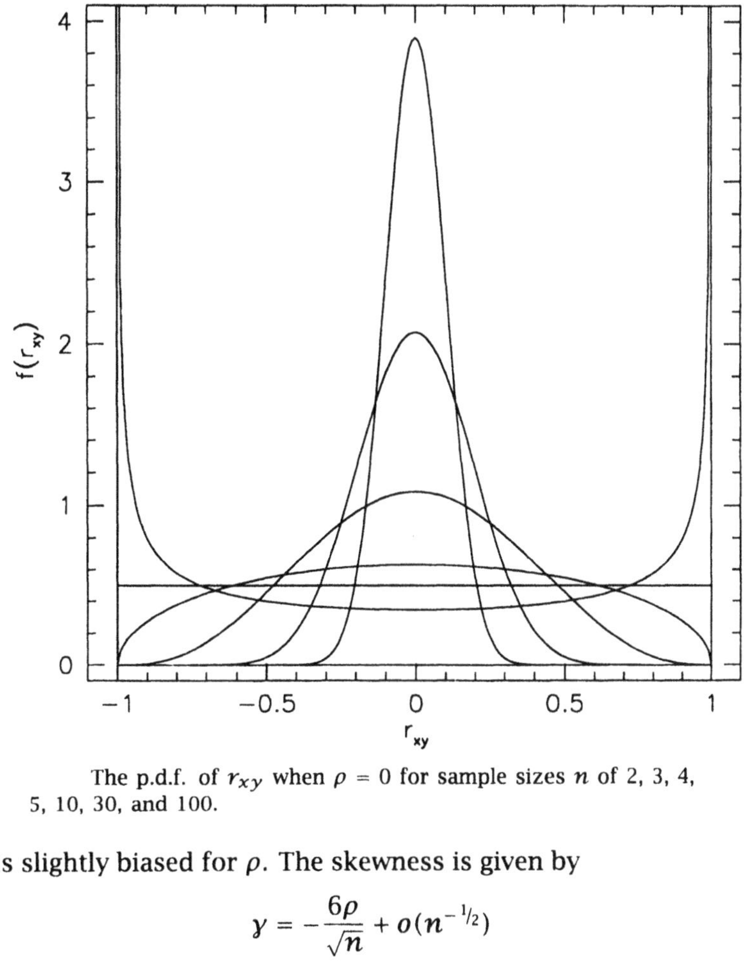

The p.d.f. of r_{xy} when $\rho = 0$ for sample sizes n of 2, 3, 4, 5, 10, 30, and 100.

so r is slightly biased for ρ. The skewness is given by

$$\gamma = -\frac{6\rho}{\sqrt{n}} + o(n^{-1/2})$$

and only approaches zero very slowly. A rule of thumb is that the distribution of r should not be treated as a Gaussian for $n < 500$ (note that this is for a Gaussian parent population; in reality we don't even know this). If you want to test the significance level of your r you should first use *Fisher's transformation,*[19,20]

$$z = \frac{1}{2} \ln \frac{1+r}{1-r},$$

i.e.,

$$r = \tanh z,$$

and equivalently $\rho = \tanh\zeta$. After carrying out this transformation, z is approximately Gaussian, with

$$\mu_1' = \zeta + \frac{\rho}{2(n-1)}\left(1 + \frac{5+\rho^2}{4(n-1)}\right) + O(n^{-3})$$

and

$$\mu_2 = \frac{1}{n-1} + \frac{4-\rho^2}{2(n-1)^2} + O(n^{-3}).$$

For large n and small ρ these become mean ζ, variance $1/(n-3)$. Fisher's transformation is an example of a *variance stabilizing* transformation, as the variance of z is almost independent of the (unknown) value of ρ. This result is usable for $n > 50$ (or less if ρ is small), and better approximations are available if needed, for example Hotelling's z^*.[21]

Tests based upon variances (and thus those based on the ratio of χ^2 variates) are usually sensitive to departures of the parent population from normality.[22] It may be shown (using methods similar to those employed in section 15.2) that tests based upon the correlation coefficient r are not too badly affected, even for n as small as 15, if x and y are independent.[23] If on the other hand $|\rho|$ is large, departures from a theory based on normality are significant. As we shall see in the next section it is in fact possible to estimate the significance of correlations without making any assumptions about the parent population.

References

1: *K&S* 26.23
2: *NR* 2.9
3: *K&S* 26.23
4: *K&S* 26.23, 27.27
5: *Bevington* 10.2
6: *Bevington II* 11.4
7: *K&S* 26.23
8: *K&S* 24.28
9: *K&S* 26.9
10: *Bevington* 7.1
11: *Bevington II* 11.2
12: *K&S* 26.23
13: *Sachs* 5.5.1
14: *K&S* 16.28
15: *Bevington II* 11.2
16: *K&S* 10.9
17: *K&S* 16.32
18: *K&S* 16.32, 26.15
19: *K&S* 16.33
20: *Sachs* 5.4.5, 5.5.1
21: *K&S* Exc. 16.19
22: *K&S* 31.6
23: *K&S* 31.19

13. Rank Correlation Coefficients

While discussing the significance of Pearson's correlation coefficient r_{xy} we were forced to assume that x and y followed a joint Gaussian distribution. We can avoid this assumption by not considering the values of x and y directly, but by instead concentrating on their *ranks.*

If we have a set of observations $\{x_i\}$, the rank of x_i, X_i, is the index that the ith data point would have after we had sorted the $\{x_i\}$ into ascending order.[1] The great advantage of using ranks is that we know their distribution — each integer in the range 1 to n occurs exactly once. If several of the x_i's are equal, we can define their rank to be the average of their indices, or simply assign ranks randomly to the tied observations, and later average the $n_{\text{tie}}!$ resulting statistics. For example, if $\mathbf{x} = \{0.1\ 0.5\ 0.2\ 0.3\ 0.2\ 0.15\ 0.6\}$, we could assign ranks as $\mathbf{X} = \{1\ 6\ 3.5\ 5\ 3.5\ 2\ 7\}$ or average over the statistics resulting from $\mathbf{X} = \{1\ 6\ 3\ 5\ 4\ 2\ 7\}$ and $\mathbf{X} = \{1\ 6\ 4\ 5\ 3\ 2\ 7\}$.

I shall not give a proper treatment of ties in this book, as they are rare when the data is drawn from continuous distributions and considerably complicate the analysis without adding anything significant to its content. If you have ties present in your data you should consult the literature before publishing your results.[2,3]

13.1. Spearman's Correlation Coefficient

If x and y are correlated, then it is clear that X and Y are also correlated, and we can proceed to define the *Spearman correlation coefficient* r_s by direct analogy to the Pearson coefficient:[4,5]

$$\begin{aligned} r_s &= \frac{\sum(X_i - \bar{X})(Y_i - \bar{Y})}{\left(\sum(X_i - \bar{X})^2 \sum(Y_i - \bar{Y})^2\right)^{1/2}} \\ &= \frac{\sum X_i Y_i / n - \bar{X}^2}{\sum X_i^2 / n - \bar{X}^2}, \end{aligned}$$

where I have used the fact that X and Y follow identical distributions, and thus $\bar{X} = \bar{Y}$, $V_X = V_Y$. Noting that

$$\begin{aligned} \sum(X_i - Y_i)^2 &= \sum X_i^2 + \sum Y_i^2 - 2\sum X_i Y_i \\ &= 2\left(\sum X_i^2 - \sum X_i Y_i\right) \\ &= 2n\, V_X\, (1 - r_s), \end{aligned}$$

and recollecting that

$$\sum X_i = \frac{1}{2}n(n+1)$$

and

$$\sum X_i^2 = \frac{1}{6}n(n+1)(2n+1),$$

this can be rewritten as

$$\begin{aligned} r_s &= 1 - \frac{1}{2n}\frac{\sum(X_i - Y_i)^2}{V(X)} \\ &= 1 - \frac{6}{n(n^2-1)}\sum(X_i - Y_i)^2, \end{aligned}$$

which can be a more convenient form.

13.2. Kendall's Correlation Coefficient

Let us sort one of our datasets (say y), reordering x at the same time so that corresponding data points remain paired. Then let us rank the data, which will be easy for y as its ranks will be $Y_i = \{1, 2, \cdots, n\}$. If the correlation were perfect the ranks of x would be $X_i = \{1, 2, \cdots, n\}$ too, and any departure from this ideal state can be used as a measure of the lack of correlation.[6] We define Q as the number of *inversions* in the X_i, i.e., the number of times that a larger rank appears to the left of a smaller one: for example, if $X_i = \{3214\}$ Q would be 3, namely $3-2$, $3-1$, and $2-1$. We can further define Kendall's rank correlation coefficient

$$t = 1 - \frac{4Q}{n(n-1)}.$$

The equivalent population statistic is written τ. You should not confuse Kendall's t with Student's; fortunately they are seldom used simultaneously. If the X_i are perfectly ordered, $Q = 0$ and $t = 1$; if they are ordered exactly backward, ($X_i = \{n, n-1, \cdots, 1\}$), $Q = n(n-1)/2$, and $t = -1$; and if they are ordered randomly, $t = 0$. It's clear that t is distributed in the range $[-1, 1]$ like any other well behaved correlation coefficient.

In some sense t is a purer rank statistic than r_s, as it only depends on whether one observation is greater than another and has no idea of how much greater. On the other hand, you might feel that inversions where the ranks are very different *should* contribute more to the correlation coefficient than those where the difference in ranks is small. In order to investigate this, let us define

$$h_{ij} = \begin{cases} 1 & \text{if } X_i > X_j, \\ 0 & \text{otherwise} \end{cases}$$

(remember that we have already arranged the Y_i into ascending order). Using h_{ij} we can write

$$Q = \sum_{i<j} h_{ij}$$

and proceed to define

$$V = \sum_{i<j} h_{ij}(j - i)$$

which embodies the desired weighting.[7] A short calculation will, however, reveal that we have merely reinvented Spearman's coefficient r_s (if we take ties into account this is no longer quite true).

Problem 81. Show that

$$V = \sum_i i^2 - \sum_i iX_i,$$

and hence that

$$r_s = 1 - \frac{12V}{n(n^2 - 1)}.$$

(*Hint:* add and subtract $\sum_{i>j} h_{ij}$ to the definition of V.)

The two rank correlation coefficients, Kendall's and Spearman's, are closely related, as can be seen by calculating *their* correlation coefficient, which ranges from unity (for $n = 2$) down to 0.98 (for $n = 5$), and approaches unity again for large n.[8] There are, however, certain advantages to using t, as its distribution is more easily calculated and it approaches normality faster.

13.3. Distribution of r_s When $\rho = 0$

If I wanted to test the significance of a given value of r_s, I would need to know r_s's distribution. Fortunately we are in a position to calculate it exactly under the null hypothesis that x and y are independent, as X and Y are then both independently drawn from a population uniform in $[1, n]$. In principle I could evaluate this distribution once and for all (and this has been done, at least for small n).[9]

Problem 82. Use a Monte Carlo simulation to find the distribution of r_s if x and y are uncorrelated. Write a program that uses a random number generator to create 2 independent samples of size n, and calculate r_s. Repeat the calculation many times and thus arrive at an approximate p.d.f. for r_s. Use the theory of the distribution of the percentiles of a distribution (which you found in problem 37) to find how many simulations you would have to run to find the 97.5^{th} percentile to an accuracy of 5%. What value of $|r_s|$ would be significantly different from 0 (at the 95% level) for a sample of size 10? How does your value compare with the value computed from the approximate theory of the next paragraph?

As an alternative to using exact theory or a computer I can try to derive an approximation to the distribution of r_s, for example by calculating its moments.[10] By symmetry, all of r_s's odd moments vanish. I can calculate its variance:

$$\begin{aligned} V(r_s) &= \left\langle r_s^2 \right\rangle \\ &= \frac{1}{V(X)^2 n^2} \left\langle \left(\sum (X_i - \bar{X})(Y_i - \bar{Y})\right)^2 \right\rangle n^2 \end{aligned}$$

(the variance of X is just a number, $(n^2 - 1)/12$). Let us switch to variables with vanishing means to simplify the algebra, an operation that will leave the variances and r_s invariant:

$$\begin{aligned} &= \frac{1}{V(X)^2 n^2} \left\langle \sum_i (X_i Y_i)^2 + \sum_{i \neq j} X_i X_j Y_i Y_j \right\rangle \\ &= \frac{1}{V(X)^2 n^2} \left(\sum_i \left\langle X_i^2 \right\rangle \left\langle Y_i^2 \right\rangle + \sum_{i \neq j} \left\langle X_i X_j \right\rangle \left\langle Y_i Y_j \right\rangle \right) \\ &= \frac{1}{V(X)^2 n^2} \left(nV(X)^2 + n(n-1) \left\langle X_i X_j \right\rangle^2 \right) \\ &= \frac{1}{V(X)^2 n^2} \left(nV(X)^2 + n(n-1) \left(\frac{-V(X)}{n-1} \right)^2 \right) \\ &= \frac{1}{V(X)^2 n^2} \left(\frac{n^2 V(X)^2}{n-1} \right) \\ &= \frac{1}{n-1} \end{aligned}$$

(note that $\sum X_i = n(n+1)/2$ is a constant, so $n^2(\bar{X})^2 = (\sum X)^2 = \langle(\sum X)^2\rangle = \langle\sum X_i^2 + \sum_{i \neq j} X_i X_j\rangle = n(\sigma_X^2 + \bar{X}^2) + n(n-1)\langle X_i X_j\rangle_{i \neq j}$, and therefore $\langle X_i X_j\rangle_{i \neq j} = ((n-1)\bar{X}^2 - \sigma_X^2)/(n-1)$).

A similar calculation can be used to evaluate the fourth moment, and these moments can be used to construct an approximate distribution function for r_s, namely

$$dF = \frac{1}{B((n-2)/2, 1/2)}(1 - r_s^2)^{(n-4)/2}\, dr_s,$$

which is adequate for $n > 30$ and test size $\alpha > 0.0005$.[11] It may be more convenient computationally to note that (in this approximation)

$$t = \left(\frac{(n-2)r_s^2}{(1-r_s^2)}\right)^{1/2}$$

is distributed as Student's t with $n - 2$ degrees of freedom.

You will doubtless have realized that this distribution for r_s is identical to the one that we found for r_{xy}, but with the difference that we have made no assumptions about the Gaussian nature of the parent population.

Spearman's correlation coefficient can only take on a finite number of values, while we have approximated its distribution by a continuous distribution, Student's t. This is a potential source of trouble, which can be at least partially avoided by applying *continuity corrections* which increase the accuracy of the approximations for small n.[12]

13.4. Distribution of Kendall's t When $\tau = 0$

The distribution of t, under the null hypothesis of independence ($\rho = \tau = 0$), may similarly be calculated[13] and is found to be approximately Gaussian with $\mu_t = 0$ and

$$\sigma_t^2 = \frac{2(2n+5)}{9n(n-1)}.$$

The approach to normality is extremely rapid and may be used for $n > 10$, although for tests based on the wings of the distribution, n should probably exceed 30;[14] for small n the exact distribution is readily available in appendices and statistical tables,[15] and is in any case easily found by means of Monte Carlo calculations. As for r_s, continuity corrections may be used to improve the accuracy of the small-n approximations.[16]

13.5. Rank Correlation Coefficients When $\rho \neq 0$

We now know how to find whether a given value of r_s or t differs significantly from zero, but we do not know how to place confidence limits on a measured correlation coefficient. What we need to know is the distribution of the correlation coefficients when the null hypothesis $\rho = 0$ is rejected, and the previous sections offer no succor.

It turns out that $\langle t \rangle = \tau$, so t is unbiased for τ, but the analogous statement doesn't hold for r_s; for large n we have[17]

$$\langle r_s \rangle = \rho_s + \frac{3}{n+1} (\tau - \rho_s) ,$$

which at least allows us to find an unbiased estimate for ρ_s. In the large sample limit, and if ρ_s and τ are not too close to unity, r_s and t follow Gaussian distributions with means ρ_s and τ respectively, and variances obeying the inequalities

$$\sigma_{r_s}^2 \leq \frac{3}{n} \left(1 - \rho_s^2\right)$$

and

$$\sigma_t^2 \leq \frac{2}{n} \left(1 - \tau^2\right) .$$

These are, regrettably, not very strong limits. They cannot be significantly improved, and it is indeed possible to construct parent populations that almost attain them.[18] This failure is more a failure of correlation coefficients in general than of rank correlation coefficients in particular; for example, you will recollect that the variance of Pearson's correlation coefficient for a sample drawn from a bivariate Gaussian population is (for large n)

$$\sigma_{r_{xy}}^2 = \frac{1}{n} \left(1 - \rho^2\right) .$$

It is possible to place closer limits on the true value of τ by examining the ranks themselves rather than simply the value of t;[19] an alternative would be to use a bootstrap (cf. problem 39).

13.6. Efficiency of Rank Tests

In order to measure the efficiency of a test we must have something to compare it with, and we shall (following universal practice) take our comparison test to be Pearson's correlation coefficient r_{xy} for a bivariate Gaussian population. It may be shown that tests based on r_{xy} are uniformly most powerful under these conditions. It may also be shown that the asymptotic relative efficiency of r_s (and thus t) against r_{xy} is $9/\pi^2 = 0.91$. In other words, obtaining a test of the same power and critical region using r_{xy} requires you to use a sample that is 91% as large as the sample you'd need to use r_s.[20] There is little incentive to use r_{xy} instead of r_s or t, and what little incentive you might feel can be removed by the use of a clever trick:

Let us rank the data as before to obtain the X_i and Y_i, but rather than using them directly to obtain a correlation coefficient let's make a further transformation to variables X_i' and Y_i', where Z_i' is the expected value of the i/n^{th} percentile of an n point sample drawn from an $N(0,1)$ distribution. $\mathbf{X}'$ and $\mathbf{Y}'$ are called *normal scores.*[21]

We can now proceed to calculate the (Pearson's) correlation coefficient of $\mathbf{X}'$ and $\mathbf{Y}'$. Its distribution must be identical to that of r_{xy} under the Gaussian assumption, the difference being that now we *know* that our variates are really Gaussian, because we constructed their distributions ourselves. What have we achieved? If the original $\mathbf{x}$ and $\mathbf{y}$ were really Gaussian, we have found a test that is 100% efficient relative to r_{xy}; if they are not, we have at least not fallen prey to any unjustified assumptions.

References

1: *K&S* 31.21
2: *K&S* 31.81
3: *K&G* 3.2
4: *K&S* 31.22
5: *Sachs* 5.3.1
6: *K&S* 31.24
7: *K&S* 32.27
8: *K&S* 31.28
9: *K&G* Table 2
10: *K&S* 31.23
11: *K&S* 31.23
12: *K&G* 4.15
13: *K&S* 31.25
14: *K&S* 31.26
15: *K&G* Table 1
16: *K&G* 4.8
17: *K&G* 5.27
18: *K&G* 5.23
19: *K&G* 5.24
20: *K&S* 31.31
21: *K&S* 31.39

14. Tests of Fit

Testing hypotheses such as $H_0 : F(x) = F_0(x)$ is known as the *goodness-of-fit* problem. We shall discuss two approaches to this problem: the χ^2 (or Pearson) test, and the *Kolmogorov-Smirnov* (KS) *test.*

14.1. Binned Data: χ^2 Tests

We have already met the χ^2 test in judging whether a particular linear model was a good fit to data. In that case we assumed that the errors on individual data points were Gaussian, and proceeded to form the sum

$$X^2 = \sum \frac{(y_i - f(\mathbf{x}_i : \theta))^2}{\sigma_i^2},$$

which was distributed as χ^2 with $n - s$ degrees of freedom (where n is the number of data points, and s the number of parameters fitted). In this section we shall meet a similar test in a different context.

Constructing a histogram from a set of data points is known as *binning*,† and is a common way of visualising data. In general, binning leads to a loss of information, as we have replaced many data points with a few integers, the number of data points that lie in each bin.[1] Another common reason for binning data is to allow us to apply the χ^2 test to which we now turn.

Let the total number of sample points be n as usual, and let the predicted probability of a point falling into the ith of k bins be p_{0i}. We can construct a test by noting that each of the n_i is a Poisson variable, or (if n_i is large) approximately an $N(n_i, n_i)$ variable. If the n_i were independent, then[2]

$$X^2 = \sum \frac{(n_i - np_{0i})^2}{np_{0i}}$$

would follow a χ_k^2 distribution; as the n_i are not independent (as $\sum n_i = n$), the familiar arguments about sums of Gaussian variables reveal that X^2 is in fact a χ_{k-1}^2 variable. We can now proceed to carry out a χ^2 test in exactly the same way as we did while testing linear models with Gaussian errors.[3,4,5] Note that in this case we have made no assumptions whatsoever about the underlying distribution.

---

†In some fields bins are called *classes*.

Now consider the more interesting case, when we are forced to estimate some of the parameters from the data. How should we do this?[6] One idea is to construct a likelihood function from the binned data and use ML estimators, i.e., to solve

$$\frac{\partial \ln L}{\partial \theta_i} = \sum_j n_j \frac{\partial \ln p_{0j}}{\partial \theta_i} = 0$$

for each of the s unknown θ's. This imposes s more linear constraints upon the n_i, so X^2 is still a χ^2 distribution, but now with only $k - s - 1$ degrees of freedom.

An alternative idea, when we know the values of the unbinned data points, would be to estimate the θ's by using ML methods on the unbinned data. This will in general give more efficient estimators, but has the disadvantage that the distribution of X^2 is no longer χ^2. It can be shown that the distribution does lie between χ^2 distributions with $k - 1$ and $k - s - 1$ degrees of freedom,[7] so as k becomes large the uncertainty in the distribution of X^2 becomes unimportant. For small n it is as well to use both χ^2_{k-1} and χ^2_{k-s-1} in performing tests. If you use some other class of estimators (such as moments), the distribution is no longer bounded above by χ^2_{k-1}.[8]

Problem 83. Why is it reasonable that the distribution should lie between χ^2_{k-1} and χ^2_{k-s-1}?

We can derive a test that is asymptotically equivalent to this χ^2 test by using likelihood ratios;[9] if you are not interested in LR tests you can skip the next two paragraphs. The probability that there are n_i points in the ith bin is obviously given by

$$\begin{aligned} P(n_1, \cdots, n_k) &= \prod_i \frac{(np_{0i})^{n_i} e^{-np_{0i}}}{n_i!} \\ &= n^n e^{-n} \prod_i \frac{p_{0i}^{n_i}}{n_i!}, \end{aligned}$$

as the number in each bin will follow a Poisson distribution.

The ML estimators of the p_{0i} are given by maximizing $\ln L$ as usual, subject to the condition that $\sum p_i = 1$. Introducing, and then evaluating, a Lagrange multiplier shows that

$$\hat{p}_{0i} = n_i/n.$$

We can construct a likelihood ratio test in the usual way:

$$\begin{aligned}\ell &= \frac{L(x|\hat{\hat{\theta}}_s)}{L(x|\hat{\theta})} \\ &= n^n \prod_i \left(\frac{p_{0i}}{n_i}\right)^{n_i}.\end{aligned}$$

As mentioned before, the distribution of $-2\ln\ell$ is known to be asymptotically χ^2 with $k-1$ degrees of freedom.

I have not said anything about how you should choose your bins. For a discrete distribution such as a Poisson, the choice is usually straightforward, but continuous distributions have no natural boundaries. One obvious constraint is that the number of bins must be small enough that the approximation of the n_i's distribution as Gaussian is valid.

It is common practice to choose the bin boundaries after obtaining the data, and you might reasonably worry that this would affect X^2's distribution.[10] In fact, it may be shown that, for large samples, there is no such effect;[11] when you realize that our discussion of X^2's distribution was valid for *any* choice of bins this may not seem unreasonable.

One natural way to choose the bins is to ensure that they are equally probable; in this case there are theoretical results available to guide you in your choice of the number of bins to use.[12]

14.2. Signs Tests

Because the χ^2 test that we have just discussed is based on the squares of $d_i \equiv n_i - np_{0i}$ rather than on the d_i directly, it is insensitive to large-scale discrepancies between the data and the model; for example, if the model's mean were too large, the d_i would first be mostly positive, and then mostly negative (assuming that the model is unimodal).[13]

> **Problem 84.** What pattern would the d_i show if the mean were correct but the model had too small a variance?

We can try to test for this type of mismatch by considering the probability of seeing runs of many positive or negative residuals together. It may be shown[14] that X^2 and statistics based on these runs of the d_i are asymptotically independent. For simple hypotheses, all patterns of signs are equally probable, so a test can easily be constructed;[15] unfortunately, for the more interesting case of composite hypotheses, all patterns of signs are no longer equally likely, and we are therefore unable to invent a test.

14.3. Unbinned Data: Smirnov and Kolmogorov-Smirnov Tests

The χ^2 tests that we have just discussed were based upon the use of binned data, as we were then able to use the Poisson distribution to construct a test. It is always possible to bin data, but doing so results in a loss of information, and thus power. The Kolmogorov-Smirnov test avoids binning by throwing away a different piece of information, but it turns out that the loss of power is not nearly as great.

Consider the problem of testing whether a sample $\{x_i\}$ of size n comes from some probability distribution f. We can draw up a graph of the *cumulative* probability distribution of the sample and of f; each will be a non-decreasing curve that starts at zero at the left-hand side of the graph and reaches one at the right (your graph paper may have to stretch from $-\infty$ to ∞). If the sample were drawn from f, the curves would coincide (or, at least, their expectation values would coincide).[16] It's clear that we can use a measure of how much the curves differ to estimate whether the sample really is drawn from the given population, but what measure should we use? As a matter of notation, let us define

$$S_n(x) = \begin{cases} 0 & x < x_{(1)}, \\ \frac{r}{n} & x_{(r)} \le x < x_{(r+1)}, \\ 1 & x_{(n)} \le x, \end{cases}$$

where $x_{(r)}$ is the rth order statistic of x (so $x_{(n/2)}$ is the median of the sample). A graph of $S_n(x)$ against x is a staircase with a step of height $1/N$ at each x_i. This S_n is the cumulative distribution of the p.d.f. f^* that we introduced while discussing bootstrap estimates in section 6.5. As usual, the cumulative distribution of f is F, so our graph is a graph of S_n and F.

An obvious first choice would be the area between the two curves, or more conveniently its square,

$$\int_{-\infty}^{\infty} [S_n(x) - F(x)]^2 \, dx,$$

but this depends on the values of the sample points in a way that depends on their distribution, and this would require us to calculate the distribution of our statistic afresh every time that we changed f. An alternative (proposed by Smirnov) is the statistic

$$W^2 \equiv \int_{-\infty}^{\infty} [S_n(x) - F(x)]^2 f(x)\, dx.$$

This is equivalent to transforming x to follow the rectangular distribution in $[0,1]$, and when seen in this light it is clear that its distribution doesn't depend on the distribution of the x_i; it is said to be *distribution free.*[17] The Smirnov statistic W^2 is not widely used; I discuss it here because its theory is similar to, but simpler than, that of the Kolmogorov-Smirnov statistic to which we shall turn next.

We may calculate $\langle W^2 \rangle$ easily enough.[18] Each of our sample points has a probability $F(x)$ of being less than x, and a probability $1 - F(x)$ of being greater. The probability that r members of our sample lie below x is then given by the binomial distribution as

$$\binom{n}{r} F(x)^r (1 - F(x))^{n-r},$$

and we know that

$$\langle r \rangle = nF(x)$$

and

$$\left\langle (r - \langle r \rangle)^2 \right\rangle = nF(x)(1 - F(x)),$$

so we see that

$$\begin{aligned}
\left\langle (S_n(x) - F(x))^2 \right\rangle &= \left\langle (r/n - F(x))^2 \right\rangle \\
&= \frac{1}{n^2}\langle r^2 \rangle - \frac{2F(x)}{n}\langle r \rangle + F(x)^2 \\
&= \frac{1}{n^2}\left(nF(x)(1 - F(x)) + n^2F(x)^2\right) - \frac{2F(x)}{n} nF(x) + F(x)^2 \\
&= \frac{F(x)(1 - F(x))}{n}
\end{aligned}$$

and

$$\begin{aligned}
\left\langle W^2 \right\rangle &= \int_{-\infty}^{\infty} \left\langle (S_n(x) - F(x))^2 \right\rangle f\, dx \\
&= \frac{1}{n}\int_0^1 F(x)(1 - F(x))\, dF \\
&= \frac{1}{6n},
\end{aligned}$$

which confirms our feeling that W^2 should approach zero for large enough samples. In a similar way Smirnov was able to calculate the higher moments of the distribution of W^2, and also to obtain an asymptotic form for its distribution function.

14.4. The Kolmogorov-Smirnov Test

Kolmogorov defined a different statistic based on S_n and F, namely,[19]

$$D_n \equiv \text{Max}|S_n(x) - F(x)|,$$

which has a number of pleasant properties. The appearance of the modulus signs in the definition might lead you to expect difficulties in the calculation of D_n's distribution; remarkably this fear is not in fact borne out. D_n is usually called the *Kolmogorov-Smirnov* (KS) *statistic*, at least in astronomical circles.

The first thing to notice about D_n is that it is obviously distribution free: we are at liberty to apply any continuous transformation to x without affecting its value, as distorting the horizontal axis cannot change the value of a vertical distance.[20] The second thing to notice is that it is much easier to calculate than W^2.

It is possible to calculate the exact distribution of D_n by arguments rather similar to those we used to deduce the expectation value of W^2, but the calculations are more unpleasant.[21] An alternative to using tables is the asymptotic formula

$$\lim_{n\to\infty} \Pr\left(D_n > z/\sqrt{n}\right) = 2\sum_{r=1}^{\infty}(-)^{r-1}e^{-2r^2z^2},$$

which is a series that converges reasonably rapidly. In practice, if n is greater than 20 you can use this result, especially if you are looking for results with small sizes (i.e., high significance).

We can use this result to place limits upon the possible distribution of a sample, as we know (at such-and-such a significance level) that D_n must be less than a certain value Δ;[22] i.e., the true distribution function must lie within a band of half-width Δ. It is not possible to do this with most tests of fit, as they do not relate the statistic to the sample so directly.

Sometimes you may want to place limits upon the upward (or downward) deviations from a given sample; in this case it is possible to modify the KS statistic, giving[23]

$$D_n^+ \equiv \text{Max}(S_n(x) - F(x))$$

or

$$D_n^- \equiv \text{Min}(S_n(x) - F(x)),$$

whose distributions may be calculated easily enough (or, even more easily, looked up).[24]

It is also possible to use KS tests to see whether two samples of size n_1 and n_2 are drawn from the same distribution. We define $D_{n_1 n_2}$ as $D_{n_1 n_2} \equiv \text{Max}|S_{n_1}(x) - S_{n_2}(x)|$, and the asymptotic result becomes

$$\lim_{n\to\infty} \Pr\left(D_{n_1 n_2} > z\sqrt{(n_1 + n_2)/n_1 n_2}\right) = 2\sum_{r=1}^{\infty}(-)^{r-1}e^{-2r^2z^2}$$

(i.e., n is replaced by $n_1 n_2/(n_1 + n_2)$).

Consideration of the two-sample KS test may make the one-sample test a little less mysterious. The value of D is set by the relative order of the data points from the two samples; only if a pair of points from the two samples are interchanged is D changed.

In practice we are seldom interested in testing a sample against a hypothesis such as "it is an $N(0.271828, 0.314159)$ distribution," but we are interested in testing against such hypotheses as "it is a Gaussian," where the parameters are left unspecified.[25,26] How should we proceed? The obvious solution is to estimate the parameters, and (in this case) test the hypothesis "it is an $N(\bar{x}, s^2)$ distribution" — but is this procedure correct?

The derivation of the distribution of W^2 given above, and the corresponding derivation for D_n that was only hinted at, essentially depend on transforming the distributions to the uniform distribution in $[0, 1]$ (cf. problem 19), so we can deal with n independent points drawn from a known distribution. If we have estimated parameters from the data, the n points are no longer independent, they no longer follow the uniform distribution, and the theory no longer applies. In fact, the distribution of D_n is no longer distribution free, but if the parameters are only those of scale and location (e.g., σ and μ), the distribution depends only on the *form* of the distribution rather than on the parameters to be estimated (see the next problem). In this case it is possible to calculate the distribution of D_n appropriate to testing particular hypotheses such as "it is Gaussian," but no general theory is available. You should carefully note that we are *not* assuming that we know the distribution of the x_i's errors; we are testing whether the distribution of the x_i themselves is Gaussian.

Problem 85. Let us assume that H_0 specifies the distribution completely, except for unknown location and scale parameters θ_1 and θ_2: i.e.,

$$f(x;\theta_1,\theta_2) = \frac{1}{\theta_2} f((x-\theta_1)/\theta_2),$$

and θ_1 and θ_2 are estimated by t_1 and t_2, which have the property that if $x \to \alpha + \beta x$ then $t_1 \to \alpha + \beta t_1$ and $t_2 \to \beta t_2$. Derive an expression for the distribution of

$$y_i = \int_{-\infty}^{x_i} f(x;t_1,t_2)\,dx,$$

and show that this doesn't depend upon the unknown $\boldsymbol{\theta}$. What does this reduce to if f is Gaussian, $t_1 = \bar{x}$, $\theta_1 = \mu$, and the value of $\theta_2 = \sigma$ is known in advance?

Problem 86. Calculate the distribution of D_n for testing normality when both μ and σ are estimated from the sample. Write a program to calculate $D_n\sqrt{n}$ for a sample drawn from a Gaussian dataset[27] using unbiased estimates of μ and σ. Run this enough times to estimate the 95% point of the distribution with an accuracy of 1% for a sample size of 20. Also calculate $D_n\sqrt{n}$ using the known values of μ and σ. How do the values differ?

14.5. Efficiency

How efficient is the KS test?[28] The obvious comparison test is the χ^2 test, and it is possible to show that (for a test size of 0.05 and a power 0.5) the KS test can detect deviations about half the size of those that χ^2 is sensitive to. In fact, this result is based on a rather poor bound to the power of the KS test, and we may conclude that the KS is very much more sensitive.

Another way of looking at the same result is to ask how large a sample would be needed to achieve a given size and power; asymptotically the answer is that if the sample size required for the χ^2 test is n, the KS test only requires $n^{4/5}$. As $n \to \infty$, the relative efficiency of the χ^2 test goes to zero.

References

1: *K&S* 30.20, 30.31
2: *K&S* 30.5
3: *K&S* 30.9
4: *Sachs* 4.3
5: *Bevington* 5.4
6: *K&S* 30.10
7: *K&S* 30.19
8: *K&S* 30.18
9: *K&S* 30.4
10: *K&S* 30.20
11: *K&S* 30.21
12: *K&S* 30.28
13: *K&S* 30.33
14: *K&S* 30.34
15: *K&S* Exc. 30.9
16: *K&S* 30.46
17: *K&S* 30.47
18: *K&S* 30.46
19: *K&S* 30.49
20: *K&S* 30.50
21: *K&S* 30.54
22: *K&S* 30.56
23: *K&S* 30.57
24: *K&S* 30.58
25: *K&S* 30.62
26: *Sachs* 4.4
27: *NR* 7.2
28: *K&S* 30.59, 30.60

15. Robust Tests for Means

15.1. Robustness of Tests

We have never seriously tried to estimate the gravity of the sins that we have committed by assuming normality. The question can be answered either theoretically or experimentally, the latter being the earlier and easier path followed. A summary of the results would be: tests on means are insensitive to departures from normality; tests on variances are not.

For example, we know that the mean of a sample drawn from a Gaussian distribution will follow a t_{n-1} distribution, and that as $n \to \infty$ the mean's distribution tends to $N(\mu, \sigma^2/n)$. What about a sample drawn from a non-Gaussian distribution?[1] For small n we do not know its distribution, but we do know its mean and standard deviation, and furthermore we have the assurance of the central limit theorem that its distribution is asymptotically $N(\mu, \sigma^2/n)$ — i.e., the same as in the Gaussian case. It can in fact be shown that the distribution converges to a Gaussian faster if the parent distribution is symmetric; in applications where we are using a t-test to compare two samples, the convergence is fastest when the sizes of the two samples are equal.

For tests based on variances, the crucial point is that

$$X^2 = \sum \frac{(x_i - \bar{x})^2}{\sigma^2}$$

is distributed as χ^2 if x follows a Gaussian distribution. If the distribution is in fact non-Gaussian, the various moments of X^2 are altered, including the variance.[2] So now if I try to appeal to the central limit theorem it will indeed assure me that the asymptotic limit of the distribution of X^2 is a Gaussian — but it will be a *different* Gaussian than that predicted by naïve Gaussian theory.

One solution to this problem is to find a transformation (such as Fisher's z) that converts a given distribution to normality.[3] This approach requires that we know the underlying distribution. An alternative way of proceeding is to use distribution-free statistics (such as Spearman's r_s or the KS test), and we shall now look at some that compete directly with Gaussian theory for testing if two distributions have the same mean.

15.2. Distribution-Free Procedures for Equality of Means

Consider two samples of sizes n_x and n_y, drawn from distribution functions f and g respectively, and ask if they have the same mean.[4] Let the alternative hypothesis H_1 be

$$H_1 : f(x) = g(x - \theta).$$

We would like to find a non-parametric test of $H_0 : \theta = 0$.

If we were willing to assume that the x_i and y_i are drawn from Gaussian populations, we would use a test based on

$$t = \frac{\bar{x} - \bar{y}}{\sqrt{\frac{n_x s_{\bar{x}}^2 + n_y s_{\bar{x}}^2}{n_x + n_y - 2}\left(\frac{1}{n_x} + \frac{1}{n_y}\right)}}$$

which follows a $t_{n_x + n_y - 2}$ distribution (section 9.1). What should we do if we don't want to make any assumptions about the parent populations?

Let us call the union of the two samples z and write $n \equiv n_x + n_y$; we can think of our x-sample as being n_x points chosen from the finite parent population z. If H_0 is true, then all the z_i are equivalent and there is nothing special about our choice of the x_i; this means that we can test H_0 by testing if the x_i are indeed chosen at random from among the z_i's.

It seems reasonable to construct a test based on $(\bar{x} - \bar{y})^2$ (because $\bar{z}$ is a known constant, a test based on $\bar{x} - \bar{y}$ is equivalent to one based on $\bar{x}$ or $\bar{y}$). We can divide z into two samples of size n_x and n_y in $n!/n_x!n_y!$ ways (i.e., the number of ways to n_x points out of n); for each sample we can calculate $\bar{x} - \bar{y}$ and thus derive its distribution. Because we can imagine choosing our samples by permuting the z_i and assigning the first n_x to be "x" this is known as a *permutation test.*[5]

The value of $(\bar{x} - \bar{y})^2$ depends on the units in which x and y are measured, so let us normalize by dividing by $s_{\bar{z}}^2$. If we fix $\bar{x}$ and $\bar{y}$, and don't restrict ourselves to samples drawn from z, the minimum value of $s_{\bar{z}}^2$ is achieved when all of the x_i equal $\bar{x}$ and all of the y_i equal $\bar{y}$; in this case we have

$$n s_{\bar{z}}^2 = n_x(\bar{x} - \bar{z})^2 + n_y(\bar{y} - \bar{z})^2 = \frac{n_x n_y}{n}(\bar{x} - \bar{y})^2,$$

so if we define

$$\begin{aligned} w &= \frac{n_x n_y}{n^2 s_{\bar{z}}^2}(\bar{x} - \bar{y})^2 \\ &= \frac{n_x}{n_y s_{\bar{z}}^2}(\bar{x} - \bar{z})^2 \end{aligned}$$

it will lie in the range $[0, 1]$.

Now that we have our statistic w, what is its distribution under permutation? Clearly $\bar{z}$ and s_z^2 don't depend on how we choose our two samples, so all we need to know is the distribution of $\bar{x}$. We can find it by simply listing the $n!/n_x!n_y!$ possibilities, but this quickly becomes tedious. An alternative is to approximate the distribution by one based upon its moments; for example, $\langle w \rangle = 1/(n-1)$, and the large n limit of the second moment is $\langle w^2 \rangle = 3/(n^2-1)(1 + O(k/3\nu))$, where k is z's kurtosis and ν is the smaller of n_x and n_y. This leads those versed in such matters to choose a beta distribution $B(1/2, n/2 - 1)$ to approximate the true distribution of w:[6]

$$dF = \frac{1}{B(1/2, n/2 - 1)} w^{-1/2}(1-w)^{n/2-2}\, dw$$

(it's chosen because it has the correct first and second moments, and the third is very close).

> **Problem 87.** Show that $\langle w \rangle = 1/(n-1)$, where the expectation value is, of course, taken over all possible ways of choosing the x_i and the y_i from the z_i. (*Hint:* see problem 36.)

Rather than deal with this formula directly, a quick calculation shows that

$$w \equiv \frac{1}{1 + \frac{n-2}{t^2}},$$

so a test based on w is equivalent to a test based on t; furthermore, the beta distribution that we just guessed is exactly a t-distribution with $n-2$ degrees of freedom.

15.3. Wilcoxon's *U* Statistic

We have just shown that we can use Student's t to test the difference of two means, but we already knew that. What this calculation has revealed is that a t-test is correct even for non-Gaussian parents in the large sample limit. But what about the small sample case? We know what we should do (enumerate the true distribution of w by permuting the observed values of x_i and y_i), but that's a lot of work. We can simplify the problem by considering a test based not on the values of x and y but on their ranks X and Y. In this case we *know* the values of X_i and Y_i and can find the small-sample distribution once and for all. We know that the statistic w is equivalent to using $\bar{x}$, so let's base our test upon $\sum X_i$. In fact it is traditional to use

$$U = \sum_i X_i - \tfrac{1}{2} n_x (n_x + 1)$$

whose value lies in the range $[0, n_x n_y]$. Tests based upon U are usually called *Wilcoxon tests, U-tests, Mann-Whitney tests,* or *rank-sum tests.*[7] The distribution reduces to a sum first tabulated by Euler. The large n limit is a Gaussian with mean $n_x n_y / 2$ and variance $n_x n_y (n + 1)/12$, where "large n" means that n_x and n_y are greater than about 8.

How much have we lost by using a Wilcoxon test instead of a t-test? The answer depends on the true distribution of the x's and y's. For example, if they are Gaussian the t-test is about 5% more efficient; for a general distribution, Wilcoxon's test may be better than a t-test by arbitrarily large amounts, and it may be shown that it *never* loses more than 13.6%.[8] But let us be greedy: by replacing the ranks by normal scores[9] (just as we discussed in the context of the Spearman test) we can make sure that our test (now called the *normal scores test*) is as efficient as a t-test if the distribution is really Gaussian, but, remarkably, it can also be shown that the test is at least as good as a t-test against *any* distribution.[10]

References

1: *K&S* 31.2, 31.3
2: *K&S* 31.6
3: *K&S* 31.10
4: *K&S* 31.43
5: *K&S* 31.45
6: *K&S* 31.47
7: *K&S* 31.53
8: *K&S* 31.62
9: *K&S* 31.63
10: *K&S* 31.66

Epilogue

Statistics is a big field, and one that diligent statisticians continue to enlarge. In this book I have mostly discussed the theory underlying one of the most heavily ploughed corners, the classical statistics of Fisher, Pearson, and the other masters of the first half of the twentieth century. Although this material is crucial to modern statistical practice, aspects remain controversial; in particular the debate between Bayesian and non-Bayesian statisticians shows no sign of resolution. The devotees of the classical theory point to the arbitrary nature of the choice of prior probabilities and say that, although arguments such as those of problem 45 are reasonable, they are hardly compelling; Bayesians retort that it is more natural to talk about the uncertainties of parameters than to think about the behaviour of non-existent ensembles of experiments, and that the classical theory of hypothesis testing requires us to be influenced by failing to observe values that we didn't expect to see (cf. problem 25).

Statistics and statisticians have been profoundly affected by modern computing power. Computers have allowed the relaxation of some assumptions made purely for analytical convenience, notably that parent populations are always Gaussian. Fifty years ago the only possible approaches to estimating errors in model parameters were to apply the linear-and-Gaussian theory of section 11.5, or to use the large n properties of ML estimators (i.e., to appeal to the central limit theorem). The Monte Carlo simulations and bootstraps of section 10.4 would be unthinkable.

Computers have also enlarged the statistician's repertoire. The distributions of some important test statistics would be essentially unknown were it not for cheap computing; for example, the use of KS tests for composite hypotheses would be impracticable if the distribution of D_n for any parent distribution were not available for the price of a few minutes of CPU time.

There are some areas that I have avoided, notably time series and the analysis of variance. To discuss time series would have involved an extensive detour into Fourier series that seemed inappropriate, and while the analysis of variance is reasonably self-contained, the classical theory is heavily dependent on uncongenial Gaussian assumptions. I believe that the ground covered by this book will provide a secure base for your exploration of such topics in the statistical literature, and for the application of sound statistical techniques to the real problems that you will encounter.

Some Numerical Exercises

Problem 88. I have obtained information about the salaries of theoretical and observational astronomers (albeit in an encoded form). Are the distributions of income the same (use both χ^2 and KS tests)? Are the variances equal (use an F-test)? Is there a significant difference between theorists' and observers' incomes? Answer this both by using a distribution-free test and by assuming that the salary distributions are Gaussian.

Observer	Theorist	Observer	Theorist
−1.617	−1.696	−1.741	−1.884
−1.506	−1.327	−1.509	−1.542
−1.269	−0.889	−1.269	−0.982
−1.470	−0.825	−1.137	−0.853
−0.880	−0.707	−1.379	−0.825
−0.610	−0.467	−0.213	−0.536
−1.330	0.272	−1.175	−0.219
−1.222	−1.087	−1.586	−0.594
−0.991	−0.665	−1.079	−1.208
−0.679	−0.409	0.250	−0.313

Problem 89. The data in the table give the measurement errors for stellar velocities measured in the LMC star cluster NGC1866 using the fibre system and the RGO spectrograph on the Anglo-Australian Telescope. Is the distribution Gaussian? If I measure another star and find that its measurement error is 10.016, what is the probability that it is drawn from the same parent population (i.e., that the spectrograph's characteristics have not changed)?

ϵ	ϵ	ϵ	ϵ
−7.200	−5.904	−4.320	−4.032
−3.168	−2.736	−2.160	−2.016
−1.008	−0.576	−0.432	−0.288
−0.144	0.000	0.000	0.144
0.576	1.296	2.016	2.160
2.448	2.880	3.024	3.024
3.168	3.312	4.176	7.056
7.632			

Problem 90. The following table gives measurements of globosity (g) and globularity (γ) for a sample of Galactic star clusters. Are they correlated? Specifically, calculate r_{xy} and r_s, and estimate the significance of the correlation. Also find the least-square straight line fits to the data under the assumptions that all the errors are in the globosity, that all the errors are in the globularity, and that $\sigma_g = 0.2\sigma_\gamma$. Plot a graph showing the data and the three lines.

g	γ	g	γ
−0.182	−5.174	−1.490	−5.354
−1.052	−4.812	0.851	−4.873
−0.684	−2.977	−0.698	−4.512
−1.802	−2.074	−0.010	−2.285
0.716	−0.690	−1.085	−1.021
−0.666	0.001	0.600	−0.028
−0.010	0.844	1.205	0.062
−0.531	2.830	1.159	2.017
−1.178	4.214	0.614	3.733
−0.680	4.425	0.758	4.244
1.470	6.230	1.024	5.237
0.525	7.012		

References

Bevington, P. R., *Data Reduction and Error Analysis for the Physical Sciences*, McGraw-Hill (1969). (*Bevington*)

Bevington, P. R., and D. K. Robinson, *Data Reduction and Error Analysis for the Physical Sciences (2nd edition)*, McGraw-Hill (1992). (*Bevington II*)

Efron, B., *The jackknife, the bootstrap and other resampling plans.* In *Regional Conference Series in Applied Mathematics*, no. 38. SIAM (1982). (*Efron*)

Lee, P. M., *Bayesian Statistics: An Introduction*, Oxford University Press (1989). (*Lee*)

Kendall, M. G., and A. Stuart, *The Advanced Theory of Statistics*, 4th edition, Charles Griffin (1977). (*K&S*)†

Kendall, M. G., and J. D. Gibbons, *Rank Correlation Methods*, Edward Arnold (1990). (*K&G*)

Lupton, R. H., J. E. Gunn, and R. F. Griffin, *Dynamical Studies of Globular Clusters based on Photoelectric Radial Velocities of Individual Stars and on the Observed Mass Function. II. M13 Astronomical Journal* (1987) **93**, p. 1114. (*LGG*)

Patil, V. H., *Approximations to the Behrens-Fisher Distribution*, Biometrika **52**, p. 267, (1965). (*Patil*)

Press, W. H., B. P. Flannery, S. A. Teukolsky, and W. T. Vetterling, *Numerical Recipes*, Cambridge University Press (1988). (*NR*)

Sachs, L., *Applied Statistics: A Handbook of Techniques*, Springer-Verlag (1982). (*Sachs*)

Sedgewick, R., *Algorithms in C*, Addison-Wesley (1990)

Solari, M. E., *The "Maximum Likelihood Solution" of the problem of esti-*

†The chapter numbers in the 5th edition ($K\&S_V$) have changed a little. An approximate mapping is

Edition	Chapter								
4th	1:19	20,21	22:24	25,28	29	30	31:33	34	35	—
5th	1:19	20	21:23	25:28	—	30	—	24	29	31

mating a linear functional relationship, J. R. Statist. Soc., B, **31**, p. 372. (*Solari*)

Stuart, A., and J. K. Ord, *Kendall's Advanced Theory of Statistics,* Oxford University Press (1991). (*K&S$_V$*)

Answers

Answer 1. The rotation axis is equally likely to point in any direction, so the probability is uniform in terms of solid angle. If $d\Omega$ is an element of solid angle, then

$$dF = \frac{1}{4\pi}\, d\Omega$$

and noting that $d\Omega = 2\pi \sin\theta\, d\theta$ gives the result that

$$dF(\theta) = {}^1\!/\!_2 \sin\theta\, d\theta,$$

where θ lies in the range $0, \pi$. Because we can't distinguish between orientations of θ and $\pi - \theta$ we should replace this by

$$dF(\theta) = \sin\theta\, d\theta$$

and restrict θ to lie in $0, \pi/2$. It's easy to show that $\langle \sin\theta \rangle = \pi/4 = 0.7854$, and the variance of $\sin\theta$ is $2/3 - (\pi/4)^2 = 0.0498$.

Answer 2. Let us start with the mean:

$$\begin{aligned}
\langle x \rangle &= \int_{-\infty}^{\infty} x\Phi(x)\left(1 + \alpha(x^3 - 3x)\right) dx \\
&= \alpha \int_{-\infty}^{\infty} \Phi(x)(x^4 - 3x^2)\, dx \\
&= \alpha \Big| - x^3\Phi(x) \Big|_{-\infty}^{\infty} + \int_{-\infty}^{\infty} \Phi(x)(3x^2 - 3x^2)\, dx \\
&= 0.
\end{aligned}$$

(This was, of course, the reason for the choice of $x^3 - 3x$ rather than x^3 as the cubic term in the expansion.)
The mode is given by the solution of $df/dx = 0$, and noting that

$$\frac{d\Phi}{dx} = -x\Phi$$

we have

$$-x\Phi\left(1 + \alpha(x^3 - 3x)\right) + 3\Phi\alpha(x^2 - 1) = 0.$$

If α is small the mode will be too, so neglecting the quadratic and higher terms we see that the mode is -3α.

The equation defining the median ξ is

$$\begin{aligned}
{}^1\!/\!_2 &= \int_{-\infty}^{\xi} \Phi\left(1 + \alpha(x^3 - 3x)\right) dx \\
&= \int_{-\infty}^{0} \Phi\, dx + \int_{0}^{\xi} \Phi\, dx + \alpha \int_{-\infty}^{\xi} \Phi(x^3 - 3x)\, dx,
\end{aligned}$$

i.e.,

$$0 = \int_0^{\xi} \Phi\, dx + \alpha \Big|_{-\infty}^{\xi} - x^2\Phi\Big| + \alpha \int_{-\infty}^{\xi} \Phi(2x - 3x)\, dx.$$

If α is small so will be ξ, so approximating the first integral and evaluating the second we find that

$$0 = \xi\Phi(\xi) - \alpha\xi^2\Phi(\xi) + \alpha\Phi(\xi).$$

Upon dropping the quadratic term we see that the median ξ is $-\alpha$. It now follows that

$$\frac{\text{median} - \text{mode}}{\text{mean} - \text{mode}} = \frac{-\alpha + 3\alpha}{3\alpha} = \frac{2}{3},$$

as claimed.

Answer 3. Let

$$\begin{aligned}
D &= \int_{-\infty}^{\infty} |x - d|\, dx \\
&= \int_{-\infty}^{d} (d - x)\, dx + \int_{d}^{\infty} (x - d)\, dx.
\end{aligned}$$

Differentiating with respect to d,

$$\frac{\partial D}{\partial d} = 0 = |(d - x)f|_{x=d} + \int_{-\infty}^{d} f\, dx - |(x - d)f|_{x=d} - \int_{d}^{\infty} f\, dx,$$

so

$$2\int_{-\infty}^{d} = 1,$$

which is the required result.

Answer 4. We can define $x_{1/2}$ by

$$\frac{1}{2} = e^{-(x_{1/2} - \mu)^2/2\sigma^2},$$

i.e.,

$$x_{1/2} = \mu \pm \sqrt{2\ln 2}\sigma,$$

whence the FWHM is $2\sqrt{2\ln 2}\,\sigma \sim 2.35\sigma$.

Answer 5. Let us consider the bivariate Gaussian distribution

$$g(x,y) = \frac{1}{\pi\sqrt{1-c^2}} e^{-(x^2+2cxy+y^2)}$$

and calculate its marginal distribution:

$$\begin{aligned} g'(y) &= \int_{-\infty}^{\infty} g(x,y)\,dx \\ &= \frac{1}{\pi\sqrt{1-c^2}} \int_{-\infty}^{\infty} e^{-((x+cy)^2+(1-c^2)y^2)}\,dx \\ &= \frac{1}{\sqrt{\pi(1-c^2)}} e^{-(1-c^2)y^2}. \end{aligned}$$

We then see that

$$\begin{aligned} f(x,y) &= \frac{g(x,y)+g'(x)g'(y)}{2} \\ &= \frac{1}{2\pi\sqrt{1-c^2}} e^{-(x^2+2cxy+y^2)} + \frac{1}{2\pi(1-c^2)} e^{-(1-c^2)(x^2+y^2)} \end{aligned}$$

has two Gaussian marginal distributions, although it is not a bivariate Gaussian distribution unless $c = 0$, in which case x and y are independent.

Answer 6. Let $z = x_1 + x_2$, where x_1 is a Poisson variable with mean μ and x_2 is a Poisson variable with mean $n\mu$.

$$\begin{aligned} \Pr(z) &= \sum_{x=0}^{z} p_{x_1}(x)p_{x_2}(z-x) \\ &= e^{-(1+n)\mu} \sum_{x=0}^{z} \frac{\mu^x}{x!} \frac{(n\mu)^{z-x}}{(z-x)!} \\ &= e^{-(1+n)\mu}\mu^z \sum_{x=0}^{z} \frac{n^{z-x}}{x!(z-x)!} \\ &= \frac{e^{-(1+n)\mu}\mu^z}{z!} \sum_{x=0}^{z} \frac{z!}{x!(z-x)!} n^{z-x} \\ &= \frac{e^{-(1+n)\mu}\mu^z}{z!}(1+n)^z \\ &= \frac{e^{-(1+n)\mu}((1+n)\mu)^z}{z!}, \end{aligned}$$

which is a Poisson distribution with mean $(1+n)\mu$. It is now obvious by induction that the distribution of $\sum_{i=0}^{n} x_i$ is a Poisson with mean $n\mu$.

The characteristic function of x is $\exp(\mu(e^{it}-1))$, so the characteristic function of $\sum_{i=0}^{n} x_i$ is $\exp(n\mu(e^{it}-1))$, which is the c.f. of a Poisson process of mean $n\mu$.

Answer 7. Consider

$$\begin{aligned}\int_{\mu}^{\infty} t^n e^{-t}\,dt &= \int_0^{\infty} (t+\mu)^n e^{-(t+\mu)}\,dt \\ &= \sum_{r=0}^{n} e^{-\mu}\mu^r \frac{n!}{r!(n-r)!}\int_0^{\infty} t^{n-r}e^{-t}\,dt \\ &= n!\sum_{r=0}^{n} e^{-\mu}\mu^r\frac{1}{r!},\end{aligned}$$

from which the result follows directly.

Answer 8. The hint more-or-less says, "it's a Poisson Process," as indeed it is. If the mean number of photons falling on my detector were N, then its standard deviation would be $N^{1/2}$, so if I assume that the number of photons detected n is equal to the mean number N then the standard deviation in my measurement is $n^{1/2}$. For a more formal justification of the propriety of taking $N = n$, see problem 47.

CCDs in fact only detect about half of the incident photons, but this doesn't affect the argument.

Answer 9. Let the probability of the cat not bothering her for t minutes be $p(t)$, then

$$p(t+dt) = p(t)(1 - dt/\tau),$$

whence it follows that

$$p(t) = \frac{1}{\tau}e^{-t/\tau}.$$

The characteristic function is $(1 - it/\tau)^{-1}$; the mean and standard deviation are both τ.

Answer 10. The probability of getting some specific pattern of r A's and $n-r$ B's, ABABBABAABBA$\cdots$BB say, is $p^r q^{n-r}$, but as I don't care which order the A's and B's come in, any sequence with r A's and $n-r$ B's will do; we must therefore multiply this probability by the number of acceptable sequences. There are a total of $n!$ sequences, but only $n!/r!(n-r)!$ of them are distinct, so the desired probability is indeed

$$\binom{n}{r} p^r q^{n-r}.$$

Answer 11.

$$\begin{aligned}
\mu_1' = \langle r \rangle &= \sum_{r=0}^{n} \frac{n!}{r!(n-r)!} r p^r q^{n-r} \\
&= np \sum_{r=1}^{n} \frac{(n-1)!}{(r-1)!(n-r)!} p^{r-1} q^{n-r};
\end{aligned}$$

letting $s = r - 1$ and $m = n - 1$,

$$\begin{aligned}
&= np \sum_{s=0}^{m} \frac{m!}{s!(m-s)!} p^s q^{m-r} \\
&= np.
\end{aligned}$$

Similarly,

$$\begin{aligned}
\mu_2' = \left\langle r^2 \right\rangle &= \sum_{r=0}^{n} \frac{n!}{r!(n-r)!} r^2 p^r q^{n-r} \\
&= np \sum_{r=1}^{n} \frac{(n-1)!}{(r-1)!(n-r)!} p^{r-1} q^{n-r} \\
&= np \sum_{s=0}^{m} \frac{m!}{s!(m-s)!} (s+1) p^s q^{m-r} \\
&= np(1 + \langle s \rangle) \\
&= np(1 + (n-1)p) \\
&= n^2 p^2 + npq,
\end{aligned}$$

so

$$\mu_2 = npq.$$

Answer 12. If the dog goes right r times and left $n - r$ times, it has travelled a net distance of $n - 2r$ to the left, so

$$p(n-2r) = \binom{n}{r} p^r q^{n-r} = 2^{-n} \binom{n}{r}.$$

The characteristic function is thus

$$\begin{aligned}\phi(t) &= 2^{-n}\sum_i \binom{n}{r} e^{int} e^{-2irt} \\ &= 2^{-n} e^{int}\left(1+e^{-2it}\right)^n \\ &= \cos^n t.\end{aligned}$$

Differentiating twice shows that

$$\begin{aligned}\phi'(t) &= -n \sin t \cos^{n-1} t \\ \phi''(t) &= -n \cos^n t + n(n-1)\sin^2 t \cos^{n-2} t,\end{aligned}$$

so $\mu_1' = 0$ and $\mu_2' = \mu_2 = n$. The mean distance travelled is zero, but the standard deviation (r.m.s.) is $n^{1/2}$.

The result $\mu = 0$ is of course obvious from the symmetry of the problem; on average there will be an equal number of steps left and right, which cancel out. The squares of the lengths of the steps can show no such cancellation, however, but simply add up so that the r.m.s. distance from the starting point is $n^{1/2}$.

Answer 13. The distribution is clearly Poisson, with $\mu = 2$. The probability of finding 5 moose is

$$\frac{e^{-2}2^5}{5!} = 0.0361.$$

The probability of finding 5 or more is $1 - \Pr(4 \text{ or less})$, or 0.0526. If we approximate the Poisson by a Gaussian $N(2,2)$ the probability is $\Pr(x - \mu > (5-2)/\sqrt{5}\sigma)$, or 0.0169. The use of a Gaussian approximation for this small a value of μ is unreliable and overestimates the significance of the result. If I have visited a total of 20 lakes, the probability of finding no 5.3% results is given by the binomial distribution as $1 - (1 - 0.053)^{20}$, or 0.66, so it isn't at all surprising to have found 5 moose together.

Answer 14. Of course not. If the total number of flowers were indeterminate then this would have been a Poisson problem, but in fact it is a binomial. Even if it were Poisson he should have used the *expected* number of blue plants, rather than the observed number, to estimate the variance, in which case the excess would be $6/\sqrt{10}\sigma = 1.9\sigma$, not the 1.5σ that he claimed. The probability of 16 or more events when we expect 10 is 0.049.

In reality I know that I bought 20 plants, so the distribution is binomial rather than Poisson, and the probability of 16 or more blue plants is the same as the probability of 4 or fewer red ones, or 0.0059. If we approximate the distribution as Gaussian it is a $+2.68\sigma$ event, which occurs with probability 0.0037.

Answer 15. If I define $p \equiv 1 - q$, then the probability that I need to try $n + r$ pens to find n that work is

$$\binom{n+r-1}{r} p^{n-1} q^r p = \binom{n+r-1}{r} p^n q^r$$

(note that when we tried the last pen, it worked). Noting that

$$p^{-n} = (1-q)^{-n} = \sum_{r=0}^{\infty} \binom{n+r-1}{r} q^r,$$

we find that our probabilities add reassuringly to unity, and simultaneously realize why the distribution is called the negative binomial.

We can now find the characteristic function:

$$\phi(t) = \left\langle e^{irt} \right\rangle = p^n \left(1 - qe^{it}\right)^{-n},$$

from which

$$\mu_1' = nq/p,$$
$$\mu_2 = nq/p^2,$$

and $\mu_1' \le \mu_2$, as claimed.

If $n \to \infty$ at fixed μ_1' we must have $p \to 1 - \mu_1'/n$ and $q \to \mu_1'/n$, so the characteristic function becomes

$$\begin{aligned}\phi &\sim (1 - \mu_1'/n)^n (1 - \mu_1' e^{it}/n)^{-n} \\ &\to e^{\mu_1'(e^{it}-1)},\end{aligned}$$

which corresponds to a Poisson.

Answer 16. Let us find the characteristic function. Writing $\mathbf{t} = (t_1, \cdots, t_i, \cdots, t_N)$,

$$\begin{aligned}\phi(\mathbf{t}) &= \left\langle e^{i\mathbf{n}.\mathbf{t}} \right\rangle \\ &= \sum_{\sum n_i = n} \frac{n!}{\prod_i n_i!} \prod_i \left(p_i e^{it_i}\right)^{n_i} \\ &= \left(\sum_i p_i e^{it_i}\right)^n,\end{aligned}$$

and differentiating twice gives the desired moments.

This calculation isn't quite honest, as all the n_i aren't independent. To rectify this, set $t_1 = 0$ and find that

$$\phi(\mathrm{t}) = \left(p_1 + \sum_{i>1} p_i e^{it_i}\right)^n .$$

We now have no t_1 to differentiate with respect to, so either appeal to symmetry or perform calculations such as (for $i > 1$):

$$\begin{aligned}
\langle n_1 n_i \rangle &= \left\langle (n - \sum_{j>1} n_j) n_i \right\rangle \\
&= n^2 p_i - \sum_{j>1} \left(n(n-1) p_i p_j + n p_i \delta_{ij}\right) \\
&= n^2 p_i - n(n-1) p_i \sum_{j>1} p_j - n p_i \\
&= n(n-1) p_i (1 - (1 - p_1)) \\
&= n(n-1) p_1 p_i
\end{aligned}$$

as expected; the calculation of $\langle n_1 n_1 \rangle$ is similar.

Answer 17. The only difference from the t-distribution case is that we must split the range of y:

$$F_z = \frac{1}{2\pi} \int_{-\infty}^{0} e^{-y^2/2}\, dy \int_{-\infty}^{zy} e^{-x^2/2}\, dx + \frac{1}{2\pi} \int_{0}^{\infty} e^{-y^2/2}\, dy \int_{zy}^{\infty} e^{-x^2/2}\, dx,$$

or, substituting $x \to -x$ and $y \to -y$ in the first term,

$$= \frac{1}{\pi} \int_{0}^{\infty} e^{-y^2/2}\, dy \int_{-\infty}^{zy} e^{-x^2/2}\, dx.$$

We know that $f_z = dF_z/dz = dF_z/d(zy)\, d(zy)/dz$, so

$$\begin{aligned}
f_z &= \frac{1}{\pi} \int_{0}^{\infty} e^{y^2(1+z^2)/2}\, y\, dy \\
&= \frac{1}{\pi} \frac{1}{1+z^2},
\end{aligned}$$

which is indeed a Cauchy distribution with $\mu = 0$.

It's easy to extend this calculation to the case where x is $N(0,1)$ (this choice of μ_x and σ_x involves no loss of generality), but y is $N(\mu, \sigma^2)$ and μ is so much greater than σ that y is always positive. In this case you can show that

$$dF_z = \frac{1}{\sqrt{2\pi}} \frac{\mu}{(1+z^2\sigma^2)^{3/2}} e^{-\mu^2 z^2/2(1+z^2\sigma^2)}\, dz$$

and thus that $z\mu/(1+z^2\sigma^2)^{1/2}$ is an $N(0,1)$ variable. Note that z itself is not Gaussian.

Answer 18. All angles in the range $(0,\pi)$ are equally likely (the bullets fired into $[\pi, 2\pi]$ missed the wall entirely), so the probability of a bullet being shot in the range $\theta, \theta + d\theta$ is $1/\pi\, d\theta$. If the squad stood d quetzetls from the mango tree that grows against the wall, then the probability of a bullet hole at a distance x is

$$\frac{1}{\pi} \frac{1}{1+(x-d)^2},$$

which is a Cauchy distribution. I had better not use the average distance to place my wreath as the variance of d is infinite; the median or mode would be a better choice. We shall return to the choice of statistic when discussing sampling distributions.

Answer 19. If y has p.d.f. $g(y)$, then $g(y) = f(x)dx/dy$. Differentiating y's definition shows that $dy/ds = 1/f$, so y follows the uniform distribution. It is then easy to show that $\mu_1' = 1/2$ and $\mu_2 = 1/12$ either by direct integration or by using the results for the beta distribution given above.

Answer 20. If we define

$$I(z) = \int_{-\infty}^{\infty} e^{-itz}(1-it)^{-p}\, dt$$

the integrand has a pole at $it = 1$. If $z < 0$, $e^{-itz} \to 0$ as $t \to +i\infty$ and we can close the integration path with a large semicircle of radius R above the real axis (centred at the origin). Because this path encloses no poles, the integral around it must be zero; the contribution from the semicircular part goes to zero as $R \to \infty$, and we deduce that $I(z) = 0$ for $z < 0$.

Now consider the case $z > 0$, in which case we must close the path below the real axis. The contribution from the semicircle again vanishes as $R \to \infty$, but now the contour encloses a pole (the one at $t = -i$), so I

must be equal to the value of the integral taken around *any* (clockwise) path that encloses $t = -i$. Let us choose a path that is a circle about the pole, and change variables with

$$t = re^{-i\theta} - i,$$

where r is a constant, $dt = -ire^{-i\theta}\,d\theta$, and θ runs from 0 to 2π. The integral then becomes

$$\begin{aligned} I &= e^{-z}(-ir)^{1-p}\int_0^{2\pi} e^{-izre^{-i\theta}} e^{(p-1)i\theta}\,d\theta \\ &= e^{-z}(-ir)^{p-1}\int_0^{2\pi}\sum_j \frac{\left(-izre^{-i\theta}\right)^j}{j!} e^{(1-p)i\theta}\,d\theta. \end{aligned}$$

We know that the value of I can't be a function of r, so the only term that is non-zero is the term with $j = p - 1$ for which all of the r's cancel:

$$\begin{aligned} I &= e^{-z}\int_0^{2\pi}\frac{z^{p-1}}{(p-1)!}\,d\theta \\ &= 2\pi e^{-z}\frac{z^{p-1}}{(p-1)!} \end{aligned}$$

as claimed.

Answer 21. The easiest way to solve this is by using the characteristic function

$$\begin{aligned} \phi &= (1-2it)^{-n/2} \\ \frac{\partial\phi}{\partial(it)} &= \left(\frac{-n}{2}\right)(-2)\,(1-2it)^{-(n+2)/2} \\ \frac{\partial^2\phi}{\partial(it)^2} &= \left(\frac{n(n+2)}{2^2}\right)(-2)^2\,(1-2it)^{-(n+4)/2}, \end{aligned}$$

so $\mu' = n$, $\mu_2' = n^2 + 2n$, and $\mu_2 = 2n$.

Standardizing has the effect of transforming the characteristic function

$$\phi \to e^{-it\mu/\sigma}\phi(t/\sigma),$$

so the standardized ϕ is, in this case,

$$e^{-it\sqrt{n/2}}\left(1 - \frac{2it}{\sqrt{2n}}\right)^{-n/2}$$

or (defining $n = 2\nu$)

$$\frac{e^{-it\sqrt{\nu}}}{(1 - it/\sqrt{\nu})^{\nu}},$$

and as $\nu \to \infty$ this becomes

$$e^{-it\sqrt{\nu}} e^{it\sqrt{\nu} - t^2/2};$$

i.e., the distribution of the standardized variable becomes $N(0,1)$, and the distribution of χ_n^2 is therefore asymptotically $N(n, 2n)$. You knew this, of course, because a χ^2 variable is the sum of many independent terms and the central limit theorem applies.

Answer 22. Let V's eigenvectors and eigenvalues be given by

$$V\boldsymbol{\ell}^{(i)} = \lambda^{(i)}\boldsymbol{\ell}^{(i)}$$

(because V is symmetric it has n independent eigenvectors). Now construct the vector with components

$$\epsilon_i = \sum_j \left(\lambda^{(j)}\right)^{1/2} \ell_i^{(j)} x_j \equiv R_{ij} x_j,$$

whence the orthonormality of the $\boldsymbol{\ell}$'s shows that

$$x_i = \left(\lambda^{(i)}\right)^{-1/2} \sum_j \ell_j^{(i)} \epsilon_j.$$

Using this expression, we see that

$$\begin{aligned}
\left\langle x_i x_j \right\rangle &= \left(\lambda^{(i)}\lambda^{(j)}\right)^{-1/2} \sum_{kl} \ell_k^{(i)} \ell_l^{(j)} \langle \epsilon_k \epsilon_l \rangle \\
&= \left(\lambda^{(i)}\lambda^{(j)}\right)^{-1/2} \sum_{kl} \ell_k^{(i)} V_{kl} \ell_l^{(j)} \\
&= \left(\lambda^{(i)}\lambda^{(j)}\right)^{-1/2} \lambda^{(j)} \sum_k \ell_k^{(i)} \ell_k^{(j)} \\
&= \delta_{ij}.
\end{aligned}$$

Let us also calculate V:

$$\begin{aligned}
V &\equiv \langle \boldsymbol{\epsilon}\boldsymbol{\epsilon}^T \rangle \\
&= R\langle \mathbf{x}\mathbf{x}^T \rangle R^T \\
&= RR^T.
\end{aligned}$$

We have now constructed the matrix R referred to in the hint, and the desired result follows trivially:

$$\begin{aligned} \boldsymbol{\epsilon}^T V^{-1} \boldsymbol{\epsilon}^T &= \mathbf{x}^T R^T V^{-1} R \mathbf{x} \\ &\equiv \mathbf{x}^T A \mathbf{x}, \end{aligned}$$

then

$$\begin{aligned} A^2 &= R^T V^{-1} R R^T V^{-1} R \\ &= R^T V^{-1} V V^{-1} R \\ &= A \end{aligned}$$

and (remembering that we can permute matrices within a trace)

$$\begin{aligned} \mathrm{Tr}(A) &= \mathrm{Tr}(R^T V^{-1} R) \\ &= \mathrm{Tr}(R R^T V^{-1}) \\ &= n. \end{aligned}$$

All of the conditions for $\boldsymbol{\epsilon}^T V^{-1} \boldsymbol{\epsilon}$ to be a χ^2_n variate are satisfied.

Answer 23. See the answer to the previous problem.

Answer 24. As before, write $A = RR^T$ and $B = SS^T$. The two forms are independent if $\langle R^T \boldsymbol{\epsilon}\boldsymbol{\epsilon}^T S \rangle = R^T V S = 0$, i.e., if $AVB = 0$.

Answer 25. If all values of t are observable, a quick integration shows that $\langle t \rangle = \tau$, so the mean of the observations, $\bar{t}$, is unbiased for τ. If the machine is unable to tell that events with $t > T$ had occurred, you'll note that

$$\begin{aligned} \langle t \rangle &= \frac{\int_0^T t e^{-t/\tau}\, dt}{\int_0^T e^{-t/\tau}\, dt} \\ &\equiv \tau g(T/\tau), \end{aligned}$$

so you can find an unbiased estimator for τ by solving the equation $\bar{t} = \tau g(T/\tau)$. If we make some other assumption about what happens to points with $t > T$ the form for g will be different; for example, if all such values are reported as T, then $g(T/\tau) = 1 - \exp(-T/\tau)$, but the procedure is the same.

When the existence of the gadget is revealed you should return to your original suggestion of estimating τ as $\bar{t}$, even though no values were actually affected; when it turns out to be broken you should revert in exasperation to solving $\bar{t} = \tau g$. If this series of events, none of which alter the actual data points, makes you unhappy, you are not alone; such machinations are one of the motivations for the Bayesian approach to statistics.

Answer 26. The variance of $\bar{x}$ is

$$\sigma_{\bar{x}}^2 = \langle \bar{x}^2 \rangle - \langle \bar{x} \rangle^2 ,$$

and we have already shown that $\langle \bar{x} \rangle = \mu$, so

$$\begin{aligned}
\sigma_{\bar{x}}^2 &= \left(\frac{1}{n}\right)^2 \left\langle \sum_i x_i \sum_j x_j \right\rangle - \mu^2 \\
&= \left(\frac{1}{n}\right)^2 \left\langle \sum_i x_i^2 + \sum_{j \neq i} x_i x_j \right\rangle - \mu^2 \\
&= \left(\frac{1}{n}\right)^2 \left(\sum_i \left\langle x_i^2 \right\rangle + \sum_{j \neq i} \left\langle x_i x_j \right\rangle \right) - \mu^2 \\
&= \left(\frac{1}{n}\right)^2 \left(n\langle x^2 \rangle + n(n-1) \langle x \rangle^2 \right) - \mu^2 \\
&= \frac{1}{n} (\langle x^2 \rangle - \langle x \rangle^2)
\end{aligned}$$

$(\mu \equiv \langle x \rangle)$, so

$$\sigma_{\bar{x}}^2 = \sigma_x^2 / n,$$

which is what was to have been demonstrated.

Answer 27. We can calculate the mean deviation's variance:

$$\begin{aligned}
\left\langle \mathrm{V}\left(\frac{1}{n}\sum |x_i - \mu|\right)\right\rangle &= \left\langle \left(\frac{1}{n}\sum |x_i - \mu|\right)^2 \right\rangle - \left\langle \frac{1}{n}\sum |x_i - \mu| \right\rangle^2 \\
&= \frac{1}{n^2} \left\langle \sum_i |x_i - \mu|^2 + \sum_{i \neq j} |x_i - \mu||x_j - \mu| \right\rangle - \langle d \rangle^2 \\
&= \frac{1}{n}\sigma^2 + \frac{n(n-1)}{n^2}\delta^2 - \delta^2 \\
&= \frac{1}{n}\left(\sigma^2 - \delta^2\right).
\end{aligned}$$

For a Gaussian,

$$\delta = \frac{2}{\sqrt{2\pi}\sigma}\int_0^\infty e^{-x^2/2\sigma^2} x\,dx$$
$$= \sqrt{\frac{2}{\pi}}\sigma,$$

so

$$\left\langle V\left(\frac{1}{n}\sum|x_i-\mu|\right)\right\rangle = \frac{\sigma^2}{n}\left(1-\frac{2}{\pi}\right).$$

If we must also estimate the mean from the sample, the calculation becomes more difficult, but we can expect that $\bar{x}$ will differ from μ by a term of order $O(\sigma/n^{1/2})$, so to $O(n^{-1})$ we should expect that

$$\left\langle V\left(\frac{1}{n}\sum|x_i-\bar{x}|\right)\right\rangle = \frac{\sigma^2}{n}\left(1-\frac{2}{\pi}\right).$$

This is confirmed by an exact calculation which leads to the result that[1]

$$\left\langle V\left(\frac{1}{n}\sum|x_i-\bar{x}|\right)\right\rangle = \frac{2\sigma^2(n-1)}{\pi n^2}\left[\frac{\pi}{2}+\sqrt{n(n-2)}-n-\arcsin\left(\frac{1}{n-1}\right)\right],$$

a formula that reduces to our formula in the large n limit.

Answer 28. We have $t = a\bar{x}$, so $\langle t\rangle = a\mu$ and t is indeed biased for $a \neq 1$. Let us calculate

$$\left\langle (t-\mu)^2\right\rangle = \left\langle (t-a\mu+(a-1)\mu)^2\right\rangle$$
$$= a^2\left\langle(\bar{x}-\mu)^2\right\rangle + (a-1)^2\mu^2$$
$$= a^2\sigma^2/n + (a-1)^2\mu^2.$$

This is minimized for $a = n\mu^2/(\sigma^2+n\mu^2)$ when it takes the value

$$\left\langle (t-\mu)^2\right\rangle = \frac{\sigma^2}{n}\frac{1}{1+\sigma^2/(n\mu^2)},$$

which is always less than σ^2/n. If $\mu = \sigma$ we have

$$\left\langle (t-\mu)^2\right\rangle = \frac{\sigma^2}{1+n}.$$

Answer 29. The characteristic function of the mean of n observations is $\phi^n(t/n)$, and as the characteristic function of a Cauchy distribution is

$$\phi(t) = e^{i\mu t-|t|}$$

the characteristic function of $\bar{x}$ is

$$\begin{aligned}\phi_{\bar{x}}(t) &= \left(e^{i\mu t/n-|t/n|}\right)^n \\ &= e^{i\mu t-|t|},\end{aligned}$$

so the distribution of the mean is the same as that of a single measurement. Its variance is infinite, and it certainly isn't consistent. The central limit theorem doesn't apply because the Cauchy distribution has no moments.

Answer 30. Because many independent factors contribute to an animal's growth it seems reasonable to apply the central limit theorem and expect the distribution of the goats' weights to be Gaussian. The same argument would apply to the sheep.

The farmer is "prosperous," so we can reasonably assume that he has large flocks, but this does not mean that we can say, "He has many animals, so the central limit theorem applies, so the distribution of his animals' weights must be Gaussian." Goats aren't sheep due to genetics rather than chance, and the central limit theorem isn't relevant; however large the farm may be, the distribution will remain bimodal.

Answer 31. It is clear that

$$P(\bar{x} < \bar{x}_0 \cap s^2 < s_0^2) = \frac{1}{2\pi}\int_{\bar{x}<\bar{x}_0\cap s^2<s_0^2} e^{-(x_1^2+x_2^2)/2}\,dx_1\,dx_2.$$

If you draw a picture of the region that we are integrating over, you'll see that it is a rectangle bounded by the three straight lines $\bar{x} = \bar{x}_0$, $s = \pm s_0$. If we rotated the coordinate axes through 45° it would be much simpler, so let us do so:

$$\begin{aligned}u &= \frac{1}{\sqrt{2}}(x_1 - x_2) \\ v &= \frac{1}{\sqrt{2}}(x_1 + x_2).\end{aligned}$$

This is a pure rotation, so the Jacobian $\partial(x_1, x_2)/\partial(u, v) = 1$ and $x_1^2+x_2^2 = u^2 + v^2$; the region to be integrated over becomes $v < \sqrt{2}\bar{x}$, $|u| < \sqrt{2}s$, so

$$\begin{aligned}P(\bar{x} < \bar{x}_0 \cap s^2 < s_0^2) &= \frac{1}{2\pi}\int_{-\sqrt{2}s}^{\sqrt{2}s} e^{-u^2/2}\,du \int_{-\infty}^{\sqrt{2}\bar{x}} e^{-v^2/2}\,dv \\ &= \frac{1}{\pi}\int_{0}^{\sqrt{2}s} e^{-u^2/2}\,du \int_{-\infty}^{\sqrt{2}\bar{x}} e^{-v^2/2}\,dv\end{aligned}$$

and

$$dF = \frac{1}{\pi} e^{-s^2} \frac{d(\sqrt{2}s)}{d(s^2)} \times e^{-\bar{x}^2} \frac{d(\sqrt{2}\bar{x})}{d\bar{x}} \, ds^2 \, d\bar{x}$$

$$= \frac{1}{\sqrt{\pi}} e^{-s^2} \frac{1}{s} \, ds^2 \times \frac{1}{\sqrt{2\pi}(1/\sqrt{2})} e^{-\bar{x}^2/(2\cdot {}^1\!/_2)} \, d\bar{x},$$

which is the product of the p.d.f.'s for s^2 and $\bar{x}$; they are indeed independent, and we have recovered a χ_1^2 and an $N(0, 1/2)$ distribution as we would expect.

It is clear that the derivation will generalize to the case of n variates. The change of variables will now be to a coordinate system where one axis points along the direction $x_1 = x_2 = \cdots = x_n$ (the equivalent of u), and the other $n - 1$ are perpendicular. The u variable will give the $\bar{x}$ term, the others will give the χ_{n-1}^2 distribution.

Answer 32. We have

$$\psi(t_1, t_2) = \int e^{it_1 \bar{x}} e^{it_2 s^2} \prod_i f(x_i) \, d^n\mathbf{x}$$

and

$$\left. \frac{\partial \psi(t_1, t_2)}{\partial t_2} \right|_{t_2=0} = i \int e^{it_1 \sum x/n} s^2 \prod_i f(x_i) \, d^n\mathbf{x}.$$

Writing $s^2 = \sum x^2/n - \bar{x}^2$ this becomes

$$\left. \frac{\partial \psi(t_1, t_2)}{\partial t_2} \right|_{t_2=0} = i \int \left(\frac{1}{n} \sum x_j^2 - \bar{x}^2 \right) \prod_i e^{it_1 x_i/n} f(x_i) \, d^n\mathbf{x}$$

$$= \frac{i}{n} \phi^{n-1} \left(\frac{t_1}{n} \right) \sum_i \int x_i^2 e^{it_1 x_i/n} f(x_i) \, dx_i - \frac{i}{n^2} \phi^{n-2} \left(\frac{t_1}{n} \right) \times$$
$$\sum_{jk} \int x_j x_k e^{it_1 x_j/n} e^{it_1 x_k/n} f(x_j) f(x_k) \, dx_j \, dx_k$$

$$= i\phi^{n-1} \left(\frac{t_1}{n} \right) \int x^2 e^{it_1 x/n} f(x) \, dx - \frac{i}{n^2} \phi^{n-2} \left(\frac{t_1}{n} \right) \times$$
$$\Big[\sum_{j \neq k} \int x_j x_k e^{it_1 x_j/n} e^{it_1 x_k/n} f(x_j) f(x_k) \, dx_j \, dx_k +$$
$$\phi \left(\frac{t_1}{n} \right) \sum_j \int x_j^2 e^{it_1 x_j/n} f(x_j) \, dx_j \Big]$$

$$= i \left(\frac{n-1}{n} \right) \phi^{n-1} \left(\frac{t_1}{n} \right) \int x^2 e^{it_1 x/n} f(x) \, dx -$$
$$\frac{i}{n^2} \phi^{n-2} \left(\frac{t_1}{n} \right) n(n-1) \left[\int x e^{it_1 x/n} f(x) \right]^2$$

$$= i\left(\frac{n-1}{n}\right)\phi^{n-2}\left(\frac{t_1}{n}\right)\left(-\phi\left(\frac{t_1}{n}\right)\phi''\left(\frac{t_1}{n}\right)+\phi''^2\left(\frac{t_1}{n}\right)\right).$$

If $\bar{x}$ and s^2 are independent, $\psi(t_1, t_2) = \phi_{\bar{x}}(t_1)\phi_{s^2}(t_2)$. We know that

$$\phi_{\bar{x}}(t) = \phi^n(t/n)$$

and

$$\begin{aligned}\left.\frac{\partial \phi_{s^2}}{\partial t_2}\right|_{t_2=0} &= i\left\langle s^2\right\rangle \\ &= i\,\frac{n-1}{n}\,\sigma^2,\end{aligned}$$

so

$$\phi\phi'' - \phi'^2 = -\sigma^2\phi^2.$$

Making the substitution $p = \ln\phi$, this becomes

$$p'' = -\sigma^2,$$

so

$$\phi(t) = e^{At-\sigma^2 t^2/2},$$

which is indeed the characteristic function of a Gaussian distribution.

Answer 33. We have

$$\begin{aligned}\sum(x-\bar{x})^2 &= \sum x^2 - \frac{1}{n}\left(\sum x\right)^2 \\ &= x_\alpha\left(\delta_{\alpha\beta} - \frac{1}{n}1_\alpha 1_\beta\right)x_\beta,\end{aligned}$$

where we are assuming the summation convention, and 1_α is a vector each of whose elements is 1. We thus have

$$A_{\alpha\beta} = \delta_{\alpha\beta} - \frac{1}{n}1_\alpha 1_\beta.$$

Once we realize that $1_\alpha 1_\alpha = n$ it is easy to show that $A^2 = A$ and $\mathrm{Tr}(A) = n-1$. The x_i are independent $N(0, \sigma^2)$ variables, so the result that we proved in section 4.2 applies once we divide by σ^2; we deduce that ns^2/σ^2 is a χ^2_{n-1} variable.

Answer 34. If we define

$$I_\alpha = \int_0^\infty e^{-ns^2/2\sigma^2} s^\alpha \, ds,$$

then it's easy to show that

$$I_\alpha = \frac{1}{2}\left(\frac{\alpha-1}{2}\right)!\left(\frac{2\sigma^2}{n}\right)^{(1+\alpha)/2}.$$

If $(\alpha - 1)/2$ isn't an integer, interpret the factorial as a gamma function, i.e., define $z! \equiv \int_0^\infty t^z e^{-t}\,dt$.

In terms of I_α, s's p.d.f. is

$$\frac{1}{I_{n-2}} e^{-ns^2/2\sigma^2} s^{n-2},$$

so

$$\begin{aligned}\left\langle s^2 \right\rangle &= I_n / I_{n-2} \\ &= \frac{2\sigma^2}{n}\frac{((n-1)/2)!}{((n-3)/2)!} \\ &= \frac{n-1}{n}\sigma^2\end{aligned}$$

and

$$\begin{aligned}\left\langle s^4 \right\rangle &= I_{n+2} / I_{n-2} \\ &= \left(\frac{2\sigma^2}{n}\right)^2 \frac{((n+1)/2)!}{((n-3)/2)!} \\ &= \frac{\sigma^4(n^2-1)}{n^2},\end{aligned}$$

so

$$V(s^2) = \frac{2(n-1)\sigma^4}{n^2}.$$

Answer 35. In terms of I_α defined in the answer to the previous problem,

$$\begin{aligned}\langle s \rangle &= I_{n-1} / I_{n-2} \\ &= \left(\frac{2\sigma^2}{n}\right)^{1/2} \frac{((n-2)/2)!}{((n-3)/2)!}\end{aligned}$$

and

$$\begin{aligned}\langle s^2\rangle &= I_n/I_{n-2}\\ &= \frac{2\sigma^2}{n}\frac{((n-1)/2)!}{((n-3)/2)!}\\ &= \frac{n-1}{n}\sigma^2,\end{aligned}$$

so the variance $\langle s^2\rangle - \langle s\rangle^2$ follows immediately.

If I extend my lemma to $O(1/n)$ I see that

$$(1+x/n)^n \sim e^{x-x^2/2n} \sim e^x(1-x^2/2n),$$

and when I remember that

$$n! \sim (2\pi n)^{1/2} n^n e^{-n}\,(1+1/(12n))$$

I find that

$$\begin{aligned}\frac{(\alpha+1/2)!}{\alpha!} &\sim \frac{(\alpha+1/2)^{1/2}(\alpha+1/2)^{\alpha+1/2}e^{-\alpha-1/2}(1+1/(12(\alpha+1/2)))}{\alpha^{1/2}\alpha^{\alpha}e^{-\alpha}(1+1/(12\alpha))}\\ &= \frac{\alpha+1/2}{\alpha^{1/2}}\left(1+\frac{1}{2\alpha}\right)^{\alpha} e^{-1/2}\\ &\sim \alpha^{1/2}\left(1+\frac{1}{2\alpha}\right)e^{1/2}\left(1-\frac{1}{8\alpha}\right)e^{-1/2}\\ &\sim \alpha^{1/2}\left(1+\frac{3}{8\alpha}\right).\end{aligned}$$

With this result I can write

$$\langle s\rangle = \sigma\left(1-\frac{3}{4n}+O(n^{-2})\right),$$

and the variance becomes (to the same order)

$$\begin{aligned}V(s) &= \sigma^2\left(\frac{n-1}{n}-\left(1-\frac{3}{4n}\right)^2\right)\\ &= \frac{\sigma^2}{2n},\end{aligned}$$

in accordance with our expectations.

Answer 36. The desired variance is

$$\left\langle (\bar{x} - \langle \bar{x} \rangle)^2 \right\rangle = \left\langle \bar{x}^2 \right\rangle - \mu^2$$

(we saw that $\langle \bar{x} \rangle$ is still μ in the finite sample case)

$$= \left\langle \frac{1}{n^2} \left(\sum_{i=1}^{n} x_i^2 + \sum_{i \neq j}^{n} x_i x_j \right) \right\rangle - \mu^2$$

$$= \frac{1}{n} \left\langle x^2 \right\rangle + \frac{n-1}{n} \left\langle x_i x_j \right\rangle_{i \neq j} - \mu^2.$$

We just calculated these expectation values, and substituting we see that

$$V(\bar{x}) = \frac{N-n}{N-1} \frac{\sigma^2}{n},$$

which again reduces to the usual formula for large N; when $n = N$ we have sampled every member of the population and our value of $\bar{x}$ is exactly the population mean μ, so its variance vanishes.

Answer 37. The calculation is very similar to the analysis for the median. Writing $q \equiv 1 - p$, the p.d.f. for the pth percentile is

$$f_p(x) = F(x)^{np} f(x) (1 - F(x))^{nq}.$$

Differentiating with respect to x shows that, as expected, the maximum of f_p occurs at x_p, where $F(x_p) = p$. As before we can expand f_p in powers of ζ, finding that

$$f_p(x) \sim \left[\left(1 + \frac{\zeta f(x_p)}{p} + \frac{f'(x_p)\zeta^2}{2p} \right)^p \left(1 - \frac{\zeta f(x_p)}{q} - \frac{f'(x_p)\zeta^2}{2q} \right)^q \right]^n$$

$$\sim \left[1 - \frac{f(x_p)^2 \zeta^2}{2pq} \right]^n,$$

so ζ is asymptotically normal, with variance $pq/nf(x_p)^2$ as claimed.

Answer 38. We showed that the variance of the median is asymptotically

$$\sigma^2_{\text{median}} = \frac{1}{4 f_{\frac{1}{2}}^2 n},$$

where f_1 is the value of the distribution at the median. By symmetry it's clear that this occurs in the Cauchy distribution at $x = \mu$, and therefore $f_1 = 1/\pi$. The sampling variance is therefore

$$\sigma^2_{\text{median}} = \frac{\pi^2}{4n},$$

which vanishes as $n \to \infty$. As the expectation value of the median is clearly μ by symmetry, the median is a consistent estimator.

Answer 39. If u is $N(0, \sigma^2)$ and v is $N(0, 1)$, then

$$x = \frac{1}{\sqrt{2}}(u - v)$$

and

$$y = \frac{1}{\sqrt{2}}(u + v)$$

have correlation coefficient ρ if $\sigma^2 = (1 + \rho)/(1 - \rho)$, so we can easily write the desired program. With $\rho = 0.5$ and $n = 20$ I found $r_{xy} = 0.364$, which may seem surprisingly poor.

After generating 100 bootstrap samples by choosing with replacement from my 20 (x, y) values I found a bias of $\langle r_{xy} \rangle^* - r_{xy} = 0.022$ and a standard deviation of 0.187, which agree quite well with the theoretical results of 0.009 and 0.168; the large variance explains why $\rho_{xy} - r_{xy}$ is so large although the bias is quite small. It is worth noting that the theoretical results assume that the population is bi-Gaussian, whereas we could have applied the bootstrap analysis to any sample.

Answer 40. From the definition,

$$t_{n-1,j} = \frac{1}{n-1} \sum_{i \neq j} x_i = \frac{1}{n-1}(n\bar{x} - x_j),$$

so $\bar{t}_{n-1} = t^J_n = \bar{x}$; the mean is already unbiased so the jackknife has no effect.

Answer 41. As usual, write $\sum(x - \bar{x})^2 = \sum x^2 - n\bar{x}^2$, so

$$t_{n-1,j} = \frac{1}{n-1}\left(\sum_i x_i^2 - x_j^2\right) - \frac{1}{(n-1)^2}\left(n\bar{x} - x_j\right)^2,$$

so

$$\begin{aligned}
\bar{t}_{n-1} &= \frac{1}{n}\sum_i x_i^2 - \frac{1}{n(n-1)^2}\sum_j \left(n^2\bar{x}^2 - 2n\bar{x}x_j + x_j^2\right) \\
&= \frac{1}{n}\sum_i x_i^2 - \frac{1}{n(n-1)^2}\left(n^3\bar{x}^2 - 2n^2\bar{x}^2 + \sum_i x_i^2\right) \\
&= \frac{(n-2)}{(n-1)^2}\left(\sum_i x_i^2 - n\bar{x}^2\right) \\
&= \frac{n(n-2)}{(n-1)^2} t_n,
\end{aligned}$$

so the jackknife estimate is

$$\begin{aligned}
t_n^J &= nt_n - (n-1)\cdot\frac{n(n-2)}{(n-1)^2} t_n \\
&= \frac{n}{n-1} t_n \\
&= \frac{1}{n-1}\sum_i (x_i - \bar{x})^2 .
\end{aligned}$$

Answer 42. As before, $t_{n-1,i} = (n\bar{x} - x_i)/(n-1)$, so

$$\begin{aligned}
V^J(t_n) &\equiv \frac{n-1}{n}\sum_i \left(t_{n-1,i} - \bar{t}_{n-1}\right)^2 \\
&= \frac{n-1}{n}\left(\sum_i t_{n-1,i} - n\bar{t}_{n-1}^2\right) \\
&= \frac{1}{n(n-1)}\left(\sum_i (n^2\bar{x}^2 - 2n\bar{x}x_i + x_i^2) - n(n-1)^2\bar{x}^2\right) \\
&= \frac{1}{n(n-1)}\left(\sum_i x_i^2 - n\bar{x}^2\right) \\
&= \frac{1}{n}s^2.
\end{aligned}$$

Answer 43. See Answer 39.

Answer 44. The algebra is tedious but straightforward; I shall only provide an outline. We can write

$$t^Q = A + \frac{1}{n}\sum_i B_i(n_i - 1) + \frac{1}{2n^2}\sum_{ij} C_{ij}(n_i - 1)(n_j - 1)$$

(note that $t(\mathbf{1}) = A$), in which case

$$\langle t^Q \rangle^* = A + \frac{1}{2n^2}\left(\sum_i C_{ii} - \frac{1}{n}\sum_{ij} C_{ij}\right),$$

and the bootstrap bias estimate follows immediately. The rather messy term is

$$t^Q_{n-1,k} = A + \frac{1}{n(n-1)}\left(\sum B_i - nB_k\right) + \frac{1}{2n^2(n-1)^2}\left(\sum_{ij} C_{ij} - 2n\sum_i C_{ik} + n^2 C_{kk}\right),$$

so

$$\bar{t}^Q_{n-1} = A + \frac{1}{2n(n-1)^2}\left(\sum_i C_{ii} - \frac{1}{n}\sum_{ij} C_{ij}\right),$$

and the claimed result follows immediately.

Answer 45. The likelihood $p(\mathbf{x}|\sigma)$ is proportional to

$$\sigma^{-n} e^{-ns^2/\sigma^2},$$

so the posterior probability $p(\sigma|\mathbf{x})$ is proportional to

$$\left(\frac{s}{\sigma}\right)^n e^{-n(s/\sigma)^2} p(\sigma)$$

(s^n is a constant if $\mathbf{x}$ is fixed). Now let $\mathbf{x} \to \mathbf{x}' = \alpha\mathbf{x}$, $s \to s' = \alpha s$, and $\sigma \to \sigma' = \alpha\sigma$, so $d\sigma = d\sigma'/\alpha$, and the probability becomes

$$\left(\frac{s'}{\sigma'}\right)^n e^{-n(s'/\sigma')^2} \frac{p(\sigma'/\alpha)}{\alpha}.$$

All that we have done is rescale the data, so this must be of the same form as the original posterior probability for σ; we deduce that $p(\sigma'/\alpha)/\alpha = p(\sigma')$ and hence that $p(\sigma) = 1/\sigma$.

If we apply a similar argument to the prior distribution for μ the symmetry is $\bar{x} \to \bar{x} + \beta$, $\mu \to \mu + \beta$, and we see that the natural prior is a constant.

Answer 46. We know that

$$\langle \sigma^{\alpha} \rangle = C \int_0^{\infty} \sigma^{\alpha-n-1} e^{-n(s/\sigma)^2} \, d\sigma.$$

If we define

$$I_{\beta} = \int_0^{\infty} \xi^{\beta} e^{-\xi^2/2} \, d\xi,$$

then an integration by parts shows that

$$I_{\beta} = -(3+\beta) I_{2+\beta},$$

so we see that

$$\langle \sigma^2 \rangle = \frac{ns^2}{n-2} = \frac{1}{n-2} \sum (x-\mu)^2.$$

Answer 47. The likelihood of n is

$$L = \frac{\mu^n e^{-\mu}}{n!},$$

which is maximized (differentiating with respect to μ) when $\mu = n$.

Answer 48. If the time between buses is τ, the probability that I must wait a time t is

$$\begin{cases} 1/\tau & \text{if } t \le \tau, \\ 0 & \text{otherwise,} \end{cases}$$

so the likelihood of waiting 10 minutes is maximized if $\tau = 10$; I deduce that buses come every ten minutes, which may seem a little counterintuitive. In this case I have little choice about which statistic I should use, as the distribution $p(\tau) \sim 1/\tau$ is not normalizable and hence has neither median nor mode.

An argument closely analogous to that of problem 45 suggests that that we should use a prior of $1/\tau$, in which case the posterior probability is proportional to

$$\begin{cases} 1/\tau^2 & \text{if } t \le \tau, \\ 0 & \text{otherwise;} \end{cases}$$

and has a median of 2τ (its mode is still τ).

Although maximizing the likelihood of a sample is an intuitively appealing procedure, it can give surprising results; unfortunately it can be hard, when dealing with more complex problems, to see whether we ought to be be surprised.

Answer 49. Consulting a table of Gaussian integrals, we see that

$$\frac{1}{\sqrt{2\pi}}\int_{2.576}^{\infty} e^{-x^2/2}\,dx = 0.005,$$

so the desired confidence interval is $[\bar{x} - 2.576\sigma/\sqrt{n}, \bar{x} + 2.576\sigma/\sqrt{n}]$.

Answer 50. We know that $ns^2/(n-1)$ is an unbiased estimator of the population variance, so $s^2/(n-1)$ is an unbiased estimator of the sample mean's variance.

Answer 51. Assuming that we want a 2-sided test, and consulting the t tables for 9 degrees of freedom, we see that the desired interval is $\bar{x} \pm 2.262s/\sqrt{n-1}$.

Answer 52. We can find a pivot by using the probability transformation of problem 19; if x has p.d.f. $f(x)$, then it is obvious that

$$y = \int_{-\infty}^{x} f(x)\,dx$$

is pivotal for x, as y's distribution is uniform in $[0, 1]$ and thus depends on no parameters.

Let us define $X \equiv \sum x_i$, which follows a Poisson distribution with mean $n\mu$ as you showed in problem 6. Then

$$Y(X) = \sum_{r=0}^{X} \frac{(n\mu)^r}{r!} e^{-n\mu}$$

is pivotal for μ; it can be expressed in closed form as an incomplete gamma function (see problem 7).

Answer 53. Upon examining a table of Gaussian integrals, we see that

$$\frac{1}{\sqrt{2\pi}}\int_{1.645}^{\infty} e^{-x^2/2}\,dx = 0.05,$$

so the desired interval is either $[-\infty, \mu+1.645\sigma/\sqrt{n}]$ or $[\mu-1.645\sigma/\sqrt{n}, \infty]$.

Answer 54. Let us write the parameter to be estimated as θ and the observations as x. Then we want to know $p(\theta|x)$, and by Bayes' theorem this is proportional to $p(x|\theta)p(\theta)$. Her prior probability is

$$p(\theta) = \frac{1}{\sqrt{2\pi}\sigma_p} e^{-(\theta-\theta_p)^2/2\sigma_p^2}$$

and

$$p(x|\theta) = \frac{1}{\sqrt{2\pi}\sigma_x} e^{-(x-\theta)^2/2\sigma_x^2},$$

where $\theta_p = 80$, $\sigma_p = 15$, $x = 60$, and $\sigma_x = 10$. Multiplying the probabilities and then normalizing the result, this leads to a posterior probability $p(\theta|x)$ of

$$p(\theta|x) \propto \exp\left(-\frac{1}{2}\left[\theta^2\left(\frac{1}{\sigma_p^2} + \frac{1}{\sigma_x^2}\right) + 2\theta\left(\frac{\theta_p}{\sigma_p^2} + \frac{x}{\sigma_x^2}\right)\right]\right),$$

i.e., an

$$N\left(\frac{\theta_p/\sigma_p^2 + x/\sigma_x^2}{1/\sigma_p^2 + 1/\sigma_x^2}, \frac{1}{\sigma_p^2} + \frac{1}{\sigma_x^2}\right) = N(66.2, 8.3^2)$$

distribution. Note that, had the uncertainty in her prior probability been larger, the posterior probability would have been closer to the value given in the new paper.

Answer 55. The analysis of the last problem shows that her posterior probability for the age of the universe is $N(6000, 25)$. A statistical analysis relies on the data presented, and the tiny error associated with the second age estimate leads to its dominating the posterior probability. Our astronomer would in reality almost certainly re-examine her sources of information and conclude that there is no point applying statistical arguments to a problem rooted in theology.

Answer 56. The procedure is identical, except now we must use the t-distribution to construct the confidence interval. We know that

$$t = \frac{\bar{x} - \mu}{s/\sqrt{n-1}}$$

is a t_{n-1} variate, so all that we have to do is look up the confidence region in t-tables. It's easy to check that we have done this correctly by seeing that the tables give the familiar Gaussian values if we take the number of degrees of freedom to be infinite (e.g., a 95%, 2-tailed test corresponds to $t = \pm 1.96$).

Answer 57. If ξ and η are both Gaussian, then $\xi - \eta$ will follow a Gaussian distribution with mean $\mu_\xi - \mu_\eta$ and variance $\sigma_\xi^2 + \sigma_\eta^2$. The distribution of $\bar{x}$ is known to be $N(\mu_x, \sigma_x^2/n)$, so $\bar{x} - \bar{y}$ is a Gaussian with mean

$$\mu_x - \mu_y$$

and variance

$$\frac{\sigma_x^2}{n_x} + \frac{\sigma_y^2}{n_y}.$$

We know that $n_x s_x^2 \equiv \sum (x_i - \bar{x})^2$ is the sum of $n_x - 1$ independent Gaussian variates, so dividing by σ_x^2 gives a $\chi^2_{n_x-1}$ variable; adding the corresponding expression for y gives a χ^2 variable with $n_x + n_y - 2$ degrees of freedom, and thus

$$\frac{n_x s_x^2/\sigma_x^2 + n_y s_y^2/\sigma_y^2}{n_x + n_y - 2}$$

has an expectation value of one. If $\sigma_x = \sigma_y = \sigma$, we see that $\langle s^2 \rangle = \sigma^2$ as claimed.

Because x and y are Gaussian, $\langle \bar{x} s_x^2 \rangle = \langle \bar{y} s_y^2 \rangle = 0$. Because x and y are independent samples, $\langle \bar{x} s_y^2 \rangle = 0$, so s^2 and $\bar{x} - \bar{y}$ are independent. The archetypical t-variate with $n_x - 1$ degrees of freedom is

$$\frac{\bar{x} - \mu}{s_x/\sqrt{n-1}},$$

where we know that $\bar{x} - \mu$ follows a Gaussian distribution, s_x^2 is proportional to a $\chi^2_{n_x-1}$ variable, the numerator and the denominator are independent, and the square of the denominator is an unbiased estimate of the numerator's variance. It therefore follows that

$$\frac{\bar{x} - \bar{y}}{\sqrt{\frac{n_x s_x^2 + n_y s_y^2}{n_x + n_y - 2}\left(\frac{1}{n_x} + \frac{1}{n_y}\right)}}$$

follows a t-distribution with $n_x + n_y - 2$ degrees of freedom.

Answer 58. We know that

$$\left\langle s^2 \right\rangle = \frac{n-1}{n}\sigma^2$$

and (for a Gaussian parent)

$$V(s^2) = \frac{2(n-1)}{n^2}\sigma^4,$$

so

$$\left\langle s^2 \right\rangle = \frac{\sigma_x^2}{n_x} + \frac{\sigma_y^2}{n_y}$$

and

$$V(S^2) = 2\left(\frac{\sigma_x^2}{n_x^2(n_x - 1)} + \frac{\sigma_y^2}{n_y^2(n_y - 1)}\right).$$

If X^2 is a χ_α^2 variate, then

$$\left\langle gX^2 \right\rangle = g\alpha$$

and

$$V(gX^2) = 2g^2\alpha,$$

so $\alpha = 2\left\langle gX^2\right\rangle^2 / V(gX^2)$, or, in this case,

$$\nu = \frac{\left(\frac{\sigma_x^2}{n_x} + \frac{\sigma_y^2}{n_y}\right)^2}{\left(\frac{\sigma_x^2}{n_x^2(n_x-1)} + \frac{\sigma_y^2}{n_y^2(n_y-1)}\right)}.$$

In parallel to the argument of the answer to the previous problem, the variance of $\bar{x} - \bar{y}$ is S^2 and $\langle S(\bar{x} - \bar{y})\rangle = 0$, and S^2 is (approximately) a multiple of a χ_ν^2 variable, so $(\bar{x} - \bar{y})/S$ is (approximately) a t_ν variable. If we replace σ^2 by its unbiased estimator $ns^2/(n-1)$ in our formula for ν we will arrive at the result quoted in the text.

Answer 59. The boundary is given by

$$\frac{(1 + (x_1 - \mu_0)^2)(1 + (x_2 - \mu_0)^2)}{(1 + (x_1 - \mu_1)^2)(1 + (x_2 - \mu_1)^2)} = k_\alpha.$$

This is a quartic, and is drawn for some values of k_α in the accompanying figure.

Answer 60. If I know nothing about the parent distribution, then all samples are equally likely (if we accept Bayes' hypothesis) and $L(x_i|H_1)$ is a constant; the Neyman-Pearson lemma now simply states that $\mathcal{R}$'s boundary is a surface of constant $L(x_i|H_0)$.

For the Gaussian example this demands that we choose a surface of constant

$$\sum (x_i - \mu)^2 \equiv n\left(s^2 + (\bar{x} - \mu)^2\right)$$

as the boundary of our test region. If we decide to use $\bar{x}$ instead of the x_i's then the likelihood takes the form

$$L(\bar{x}|H_0) = \frac{\sqrt{n}}{\sqrt{2\pi}\sigma} e^{-n(\bar{x}-\mu)^2/2\sigma^2},$$

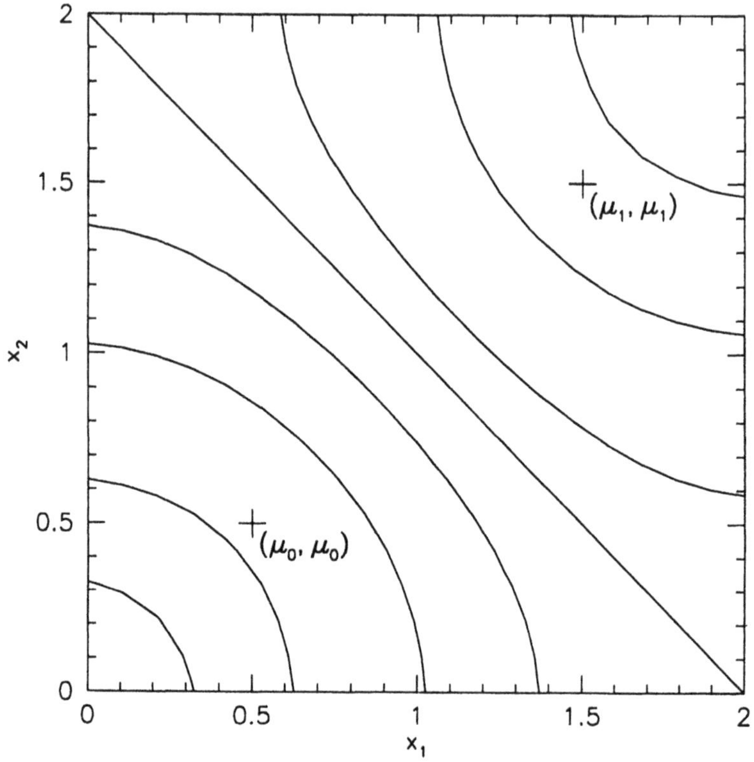

The boundary of confidence regions for a Cauchy distribution, and $k_\alpha = \{0.2\ 0.35\ 0.7\ 1\ 1.5\ 2.5\ 4.5\ 6\}$.

and Neyman and Pearson direct us to use $|\bar{x} - \mu|$ to test H_0.

Answer 61. The mean velocity of the ith star is

$$v_i = \bar{v},$$

and the velocity dispersion is given by

$$v_0 \eta_i,$$

to which we must add (in quadrature) the measurement error Δ_i. The velocity is therefore drawn from an

$$N(\bar{v}, v_0^2 \eta_i^2 + \Delta_i^2)$$

distribution. The (log-) likelihood is then given (omitting constant terms) by

$$\ln L = -1/2 \sum \ln \left(\nu_0^2 \eta_i^2 + \Delta_i^2\right) - \sum \frac{(\nu_i - \bar{\nu})^2}{2\left(\nu_0^2 \eta_i^2 + \Delta_i^2\right)},$$

and differentiating with respect to $\bar{\nu}$ and ν_0 gives the equations

$$\bar{\nu} \sum \frac{1}{\nu_0^2 \eta_i^2 + \Delta_i^2} = \sum \frac{\nu_i}{\nu_0^2 \eta_i^2 + \Delta_i^2}$$

and

$$\sum \frac{\eta_i^2}{\nu_0^2 \eta_i^2 + \Delta_i^2} = \sum \frac{(\nu_i - \bar{\nu})^2 \eta_i^2}{\left(\nu_0^2 \eta_i^2 + \Delta_i^2\right)^2}.$$

If Δ_i were zero, the first of these would reduce to the simple result $\bar{\nu} = \sum \nu_i / n$; if the measurement errors are small, the equation can still be used to invent an iterative scheme that converges rapidly.

In our study of M13 we also allowed for the fact that the cluster is rotating, and estimated the angle, ϕ, of the rotation axis. We found that our iterative scheme would sometimes converge to give a value of ϕ that was 180° wrong, illustrating the fact that such schemes will happily converge to local minima of the likelihood; in this case we simply added 180 and tried again.

Answer 62. It is easy to check that the quoted result is correct by multiplying it out. In order to derive the result, consider the matrix $(I - \alpha P)^{-1}$. We may expand this in a series, and as I and P commute there is no need to be careful to keep the matrices in the proper order:

$$(I - \alpha P)^{-1} = 1 + \alpha P + \alpha^2 P^2 + \cdots.$$

It is easily seen that $P^2 = P$, so we may write this as

$$\begin{aligned} &= 1 + P(\alpha + \alpha^2 + \cdots) \\ &= 1 - P + P(1 + \alpha + \alpha^2 + \cdots) \\ &= 1 - P + P/(1 - \alpha) \\ &= 1 + \alpha/(1 - \alpha)P, \end{aligned}$$

from which the result follows by substituting for α.

Answer 63. This is easiest to see in suffix notation, and employing the summation convention we have

$$S = (y_i - M_{ik}\theta_k)V^{-1}_{ij}(Y_j - M_{jk}\theta_k),$$

so

$$\frac{\partial S}{\partial \theta_\alpha} = -M_{ik}\delta_{k\alpha}V^{-1}_{ij}(Y_j - M_{jk}\theta_k) + (y_i - M_{ik}\theta_k)V^{-1}_{ij}(-M_{jk}\delta_{k\alpha}\theta_k),$$

i.e.,

$$(M_{i\alpha}V^{-1}_{ij} + M_{i\alpha}V^{-1}_{ji})y_j = (M_{i\alpha}M_{jk} + M_{j\alpha}M_{ik})V^{-1}_{ij}\theta_k,$$

so, as $V^{-1}_{ij} = V^{-1}_{ji}$,

$$M_{i\alpha}V^{-1}_{ij}y_j = M_{i\alpha}V^{-1}_{ij}M_{jk}\theta_k,$$

or, in matrix notation,

$$M^T V^{-1}\mathbf{y} = M^T V^{-1} M\boldsymbol{\theta},$$

which leads to the claimed result.

Answer 64. The matrix M^T is not in general square and invertible, so this expansion of the inversion of $M^T V^{-1} M$ doesn't make sense. However, for the case $n = k$, M is square, the problem is exactly constrained, and $\hat{\boldsymbol{\theta}}$ does indeed equal $M^{-1}\mathbf{y}$.

Answer 65. Because $\mathbf{t}$ is unbiased we know that $\langle \mathbf{t} \rangle = \boldsymbol{\theta}$, and from the definition of $\mathbf{t}$ we know that $\langle \mathbf{t} \rangle = \langle T\mathbf{y} \rangle = \langle T(M\boldsymbol{\theta} + \boldsymbol{\epsilon})\rangle = TM\boldsymbol{\theta}$, which must hold for any $\boldsymbol{\theta}$; we deduce that $TM = I$.

If we write $\mathbf{t}' = \mathbf{t} - \langle \mathbf{t} \rangle = T\boldsymbol{\epsilon}$ and $\hat{\boldsymbol{\theta}}' = \hat{\boldsymbol{\theta}} - \langle \hat{\boldsymbol{\theta}} \rangle = (M^T V^{-1} M)^{-1} M^T V^{-1}\boldsymbol{\epsilon}$, we have (taking the hint)

$$\mathbf{t} - \langle \mathbf{t} \rangle = \hat{\boldsymbol{\theta}}' + (\mathbf{t}' - \hat{\boldsymbol{\theta}}'),$$

so

$$V(\mathbf{t}) = \left\langle \hat{\boldsymbol{\theta}}^2 \right\rangle + \left\langle (\mathbf{t}' - \hat{\boldsymbol{\theta}}')^2 \right\rangle + 2\left\langle (\mathbf{t}' - \hat{\boldsymbol{\theta}}')\hat{\boldsymbol{\theta}}^T \right\rangle.$$

Let us consider the last term:

$$\begin{aligned}
\left\langle (\mathbf{t}' - \hat{\boldsymbol{\theta}}')\hat{\boldsymbol{\theta}}^T \right\rangle &= \langle (T - (M^T V^{-1} M)^{-1} M^T V^{-1})\boldsymbol{\epsilon}\boldsymbol{\epsilon}^T V^{-1} M (M^T V^{-1} M)^{-1} \rangle \\
&= (T - (M^T V^{-1} M)^{-1} M^T V^{-1})\, \langle \boldsymbol{\epsilon}\boldsymbol{\epsilon}^T \rangle\, V^{-1} M (M^T V^{-1} M)^{-1} \\
&= (TM - I)(M^T V^{-1} M)^{-1} \\
&= 0.
\end{aligned}$$

so

$$V(\mathbf{t}) = \left\langle \hat{\theta}^2 \right\rangle + \left\langle (\mathbf{t}' - \hat{\theta}')^2 \right\rangle .$$

Because both the terms on the right-hand side are non-negative, we can deduce that $\hat{\boldsymbol{\theta}}$ has the lowest variance of any (linear) estimator of θ.

Answer 66. Following the lead of problem 22, we can write $\epsilon = R\mathbf{x}$ with $\langle \mathbf{x}\mathbf{x}^T \rangle = \mathbf{1}$, in which case

$$\begin{aligned} X^2 &= \mathbf{x}^T R^T (I_n - V^{-1}M(M^T V^{-1} M)^{-1} M) V^{-1} \\ &\qquad\qquad (I_n - M(M^T V^{-1} M)^{-1} M^T V^{-1}) R\mathbf{x} \\ &= \mathbf{x}^T R^T (V^{-1} - V^{-1} M (M^T V^{-1} M)^{-1} M^T V^{-1}) R\mathbf{x} \\ &\equiv \mathbf{x}^T A \mathbf{x}. \end{aligned}$$

We can now calculate A^2 and $\mathrm{Tr}(A)$:

$$\begin{aligned} A^2 &= R^T (V^{-1} - V^{-1} M (M^T V^{-1} M)^{-1} M^T V^{-1}) R R^T \\ &\qquad\qquad (V^{-1} - V^{-1} M (M^T V^{-1} M)^{-1} M^T V^{-1}) R \\ &= R^T (V^{-1} - V^{-1} M (M^T V^{-1} M)^{-1} M^T V^{-1}) R \\ &= A \end{aligned}$$

(as $RR^T = V$), and

$$\begin{aligned} \mathrm{Tr}(A) &= \mathrm{Tr}(R^T (V^{-1} - V^{-1} M (M^T V^{-1} M)^{-1} M^T V^{-1}) R) \\ &= \mathrm{Tr}(R R^T (V^{-1} - V^{-1} M (M^T V^{-1} M)^{-1} M^T V^{-1})) \end{aligned}$$

(as matrices may be permuted inside a trace)

$$\begin{aligned} &= \mathrm{Tr}(\mathbf{1}_n - M (M^T V^{-1} M)^{-1} M^T V^{-1}) \\ &= \mathrm{Tr}(\mathbf{1}_n) - \mathrm{Tr}(\mathbf{1}_k) \\ &= n - k. \end{aligned}$$

So $\epsilon^T V^{-1} \epsilon$ is indeed a χ^2_{n-k} variable.

Answer 67. Consulting χ^2 tables, we find that the probability of obtaining a value as large as 46.96 from a χ^2_{27} distribution is 0.01. Because (fortuitously) 47.1 is very close to 46.96, it's easy to see that we can reject the model at the 99% level.

Answer 68. Looking back at section 11.2 and up at our formula for s^2 we see that

$$\begin{aligned} s^2 &= \frac{1}{n-2}\sum_i \left(y_i - \bar{y} - \hat{\theta}_2(x_i - \bar{x})\right)^2 \\ &= \frac{n}{n-2}\left(s_y^2 - 2\hat{\theta}_2 s_{xy} + \hat{\theta}_2^2 s_x^2\right) \\ &= \frac{n}{n-2}\left(s_y^2 - s_{xy}^2/s_x^2\right) \\ &= \frac{n}{n-2} s_y^2 (1 - r_{xy}^2) \end{aligned}$$

is an unbiased estimate of σ^2.

Answer 69. This is straightforward:

$$\begin{aligned} \left\langle \boldsymbol{\delta}(\hat{\boldsymbol{\theta}} - \boldsymbol{\theta})^T \right\rangle &\equiv \left\langle (\mathbf{y} - M\hat{\boldsymbol{\theta}})(\hat{\boldsymbol{\theta}} - \boldsymbol{\theta})^T \right\rangle \\ &= \langle (I - M(M^TM)^{-1}M^T)\boldsymbol{\epsilon}\boldsymbol{\epsilon}^T M(M^TM)^{-1} \rangle \\ &= (I - M(M^TM)^{-1}M^T)\, \langle \boldsymbol{\epsilon}\boldsymbol{\epsilon}^T \rangle\, M(M^TM)^{-1} \\ &= \sigma^2\,(M(M^TM)^{-1} - M(M^TM)^{-1}M^TM(M^TM)^{-1}) \\ &= 0. \end{aligned}$$

If the $\boldsymbol{\epsilon}$ are Gaussian, $\boldsymbol{\delta}$ and $\hat{\boldsymbol{\theta}} - \boldsymbol{\theta}$ are independent.

Answer 70. The p.d.f. is

$$dF(\theta_1, \theta_2) = \frac{\sqrt{ac - b^2}}{2\pi} \exp(-(a\theta_1^2 + 2b\theta_1\theta_2 + c\theta_2^2)/2)\, d\theta_1\, d\theta_2.$$

We can integrate over θ_2 by the usual technique of completing the square:

$$\begin{aligned} dF(\theta_1) &= \frac{\sqrt{ac - b^2}\, d\theta_1}{2\pi} \int_{-\infty}^{\infty} \exp(-(a\theta_1^2 + 2b\theta_1\theta_2 + c\theta_2^2)/2)\, d\theta_2 \\ &= \frac{\sqrt{ac - b^2}\, d\theta_1}{2\pi} \exp(-(a - b^2/c)\theta_1^2/2) \times \\ &\qquad \int_{-\infty}^{\infty} \exp(-((c\theta_2 + b\theta_1/c)^2)/2)\, d\theta_2 \\ &= \frac{\sqrt{ac - b^2}\, d\theta_1}{2\pi} \exp(-(a - b^2/c)\theta_1^2/2)\sqrt{2\pi/c} \\ &= \frac{\sqrt{a - b^2/c}\, d\theta_1}{\sqrt{2\pi}} \exp(-(a - b^2/c)\theta_1^2/2) \end{aligned}$$

as claimed.

"Applying the theory of the previous paragraph," we can easily find W:

$$W = \frac{1}{ac - b^2}\begin{pmatrix} c & -b \\ -b & a \end{pmatrix},$$

so the variance of θ_1 is $c/(ac - b^2)$, again as expected.

Answer 71. I is a Poisson variable, so its variance is $I = S + I_c f$. If $S \gg 1$ we can treat the distribution as Gaussian and the log-likelihood is

$$\ln L = -\frac{n}{2}\ln 2\pi - \frac{1}{2}\sum_i (S + I_c f(r_i;\mathbf{X})) - \frac{1}{2}\sum_i \frac{(I_i - S - I_c f(r_i;\mathbf{X}))^2}{S + I_c f(r_i;\mathbf{X})}.$$

Expanding to first order about (S_0, I_{c0}, X_0, Y_0) we find that

$$I_i \sim S_0 + I_{c0} f(r_i;\mathbf{X}_0) + (S - S_0) + (I_c - I_{c0}) f(r_i;\mathbf{X}_0) + \\ (X - X_0) I_{c0} \left.\frac{\partial f(r_i;\mathbf{X})}{\partial X_0}\right|_0 + (Y - Y_0) I_{c0} \left.\frac{\partial f(r_i;\mathbf{X})}{\partial Y_0}\right|_0.$$

Let us write $J_i \equiv I_i - S_0 - I_{c0} f(r_i;\mathbf{X}_0)$ and note that the star profile f is a function of $r_i \equiv |\mathbf{x}_i - \mathbf{X}|$, which together allow us to write our model in the form

$$\mathbf{J} = \begin{pmatrix} \vdots & \vdots & \vdots & \vdots \\ 1 & f(r_i;\mathbf{X}_0) & -I_{c0}\frac{x_i}{r_i}\left.\frac{df}{dr}\right|_0 & -I_{c0}\frac{y_i}{r_i}\left.\frac{df}{dr}\right|_0 \\ \vdots & \vdots & \vdots & \vdots \end{pmatrix} \begin{pmatrix} S - S_0 \\ I_c - I_{c0} \\ X - X_0 \\ Y - Y_0 \end{pmatrix}.$$

Covariance matrix V is diagonal with elements $V_{ii} = (S_0 + I_{c0} f(r_i;\mathbf{X}_0))^{-1}$, so we can easily calculate the covariance matrix $W = (M^T V^{-1} M)^{-1}$:

$$W^{-1} = \begin{pmatrix} \sum \frac{1}{S+I_c f} & \sum \frac{f}{S+I_c f} & -I_c \sum \frac{x/r}{S+I_c f}\frac{df}{dr} & -I_c \sum \frac{y/r}{S+I_c f}\frac{df}{dr} \\ \sum \frac{f}{S+I_c f} & \sum \frac{f^2}{S+I_c f} & -I_c \sum \frac{fx/r}{S+I_c f}\frac{df}{dr} & -I_c \sum \frac{fy/r}{S+I_c f}\frac{df}{dr} \\ -I_c \sum \frac{x/r}{S+I_c f}\frac{df}{dr} & -I_c \sum \frac{fx/r}{S+I_c f}\frac{df}{dr} & I_c^2 \sum \frac{x^2/r^2}{S+I_c f}\left(\frac{df}{dr}\right)^2 & I_c^2 \sum \frac{xy/r^2}{S+I_c f}\left(\frac{df}{dr}\right)^2 \\ -I_c \sum \frac{y/r}{S+I_c f}\frac{df}{dr} & -I_c \sum \frac{fy/r}{S+I_c f}\frac{df}{dr} & I_c^2 \sum \frac{xy/r^2}{S+I_c f}\left(\frac{df}{dr}\right)^2 & I_c^2 \sum \frac{y^2/r^2}{S+I_c f}\left(\frac{df}{dr}\right)^2 \end{pmatrix},$$

where I have dropped the subscript 0. Fortunately most of these terms vanish by symmetry and W reduces to

$$W^{-1} = \begin{pmatrix} \sum \frac{1}{S+I_c f} & \sum \frac{f}{S+I_c f} & 0 & 0 \\ \sum \frac{f}{S+I_c f} & \sum \frac{f^2}{S+I_c f} & 0 & 0 \\ 0 & 0 & \frac{I_c^2}{2} \sum \frac{1}{S+I_c f}\left(\frac{df}{dr}\right)^2 & 0 \\ 0 & 0 & 0 & \frac{I_c^2}{2} \sum \frac{1}{S+I_c f}\left(\frac{df}{dr}\right)^2 \end{pmatrix}$$

(remembering that $x^2 + y^2 = r^2$, so that sums involving x^2/r^2 or y^2/r^2 equal one half of the corresponding sums involving r^2/r^2). Inverting the matrix we finally arrive at

$$\begin{aligned} W_{SS} = \sigma_{SS}^2 &= \frac{1}{\Delta}\sum \frac{f^2}{S + I_c f} \\ W_{SI_c} = \sigma_{SI_c}^2 &= -\frac{1}{\Delta}\sum \frac{f}{S + I_c f} \\ W_{I_cI_c} = \sigma_{I_cI_c}^2 &= \frac{1}{\Delta}\sum \frac{1}{S + I_c f} \end{aligned}$$

and

$$\begin{aligned} W_{XX} = W_{YY} &= \sigma_{XX}^2 = \sigma_{YY}^2 \\ &= \frac{2}{I_c^2 \sum \frac{1}{S+I_c f}(df/dr)^2}, \end{aligned}$$

where Δ is the determinant of the upper left block:

$$\Delta \equiv \sum \frac{1}{S + I_c f} \sum \frac{f^2}{S + I_c f} - \left(\sum \frac{f}{S + I_c f}\right)^2.$$

The star occupies only a small part of the image, so f is negligible almost everywhere,

$$\sum \frac{f}{S + I_c f} \sim \sum \frac{f^2}{S + I_c f} \ll \sum \frac{1}{S + I_c f},$$

and the variances and covariances of S and I_c reduce to

$$\begin{aligned} \sigma_{SS}^2 &= \frac{1}{\sum 1/(S + I_c f)}, \\ \sigma_{SI_c}^2 &= \frac{\sum f/(S + I_c f)}{\sum 1/(S + I_c f) \sum f^2/(S + I_c f)}, \end{aligned}$$

and

$$\sigma_{I_cI_c}^2 = \frac{1}{\sum f^2/(S + I_c f)}.$$

If the star is faint ($I_c \ll S$) the variance of S is, as expected, $\sigma_{SS}^2 = S/n$, $\sigma_{SI_c}^2 = S/n \sum f / \sum f^2$, and $\sigma_{I_cI_c}^2 = S/\sum f^2$. The total number of counts is $C = I_C \sum f$, so its variance is

$$\sigma_{CC}^2 = S\frac{(\sum f)^2}{\sum f^2};$$

$\sum f^2/(\sum f)^2$ is sometimes called the effective area of the star, n_{eff}.

If the profile is Gaussian (a functional form that is convenient and not too far removed from reality), then f is given by

$$f(r) = e^{-r^2/2\alpha^2},$$

where α characterizes the size of the star's image; if we continue to assume that $I_c \ll S$ we can write

$$\sum_i f \sim \int_0^\infty e^{-r^2/2\alpha^2} 2\pi r\, dr = 2\pi\alpha^2$$

$$\sum_i f^2 \sim \int_0^\infty e^{-r^2/\alpha^2} 2\pi r\, dr = \pi\alpha^2$$

and thus

$$\sigma^2_{I_c I_c} = \frac{S}{\pi\alpha^2}$$

and

$$n_{\text{eff}} = (4\pi\alpha^2)^{-1},$$

which are not unreasonable results, as $\pi\alpha^2$ is approximately the area covered by the star. Similarly,

$$\sigma^2_{XX} = \sigma^2_{YY} = \frac{2S}{\pi I_c^2}.$$

Finally, we must ask if all this labour was justified. A typical second-order term is

$$\frac{I_c}{2}(X - X_0)^2 \frac{\partial^2 f}{\partial X^2},$$

so our desired condition is

$$\left| (X - X_0) \frac{\partial^2 f}{\partial X^2} \right| \ll \left| \frac{\partial f}{\partial X} \right|,$$

i.e.,

$$(X - X_0)(r^2 - \alpha^2) \ll r\alpha^2.$$

This must hold at all radii where the contribution to I from the star is significant, so r must be a few times α; our condition then becomes $X - X_0 \sim \sigma_{XX} \ll \alpha$, i.e., $S \ll \alpha^2 I_c^2$, or $\sigma_{I_c I_c} \ll I_c$. If we have a reasonably reliable estimate of the star's central intensity, then we are indeed justified in linearizing the model.

Answer 72. The errors ϵ and δ are independent and Gaussian, so the first result follows immediately. We know that

$$y_i - ax_i - b = \epsilon_i - a\delta_i,$$

so[2]

$$X^2 = \sum \frac{(y_i - ax_i - b)^2}{\phi_y + a^2\phi_x}$$

is a χ_n^2 variable.

Let us choose some confidence level α and find c_α such that $P(X_n^2 > c_\alpha) = 1 - \alpha$. Then the equation $X^2 = c_\alpha$ is a conic in a and b, and defines the confidence region for the regression line. If the conic turns out to be a hyperbola this region is unbounded; in favourable cases it will be an ellipse.

Answer 73. Applying the usual formula,

$$\hat{a} = \frac{1}{2s_{xy}} \left(\left(s_y^2 - \lambda s_x^2\right) \pm \sqrt{\left(s_y^2 - \lambda s_x^2\right)^2 + 4\lambda s_{xy}^2} \right),$$

so if we take the + sign for the square root, then $\hat{a}$ has the same sign as s_{xy}, which is clearly what we want. If we take the opposite sign we get a local minimum of the likelihood (if $\lambda = 1$ the two solutions are perpendicular).

Answer 74. Taking the hint, we observe that

$$s_x^2 = s_\xi^2 + \phi_x \ ; \ s_y^2 = s_\eta^2 + \phi_y \ ; \ s_{xy} = s_{\xi\eta}$$

and thus, for large n,

$$s_x^2 = \sigma_\xi^2 + \phi_x \ ; \ s_y^2 = \sigma_\eta^2 + \phi_y \ ; \ s_{xy} = \sigma_{\xi\eta}.$$

Recollecting that $\eta = a\xi + b$ and $\phi_y = \lambda\phi_x$, we see that

$$s_x^2 = \sigma_\xi^2 + \phi_x \ ; \ s_y^2 = a^2\sigma_\xi^2 + \lambda\phi_x \ ; \ s_{xy} = a\sigma_\xi^2.$$

Substituting these into equation 6 (and then solving the resulting quadratic) shows that

$$\lim_{n\to\infty} \hat{a} = a,$$

so $\hat{a}$ is indeed consistent.

We showed that

$$\hat{\phi}_x = \frac{1}{2(\lambda + \hat{a}^2)} \left(\hat{a}^2 s_x^2 - 2\hat{a}s_{xy} + s_y^2\right),$$

and substituting our large-n forms for s_x^2, s_{xy}, and s_y^2 shows that, as claimed, $\hat{\phi}_x$ tends to $\phi_x/2$ as $n \to \infty$.

Answer 75. Calculate

$$A_M A_N = R^T\left(I - V^{-1}M\,(M^TV^{-1}M)^{-1}M^T\right)V^{-1}RR^T \times \\ \left(I - V^{-1}N\,(N^TV^{-1}N)^{-1}N^T\right)V^{-1}R$$

$$= R^T\Big(V^{-1} - V^{-1}\Big(M\,(M^TV^{-1}M)^{-1}M^T + N\,(N^TV^{-1}N)^{-1}N^T - \\ M\,(M^TV^{-1}M)^{-1}M^TV^{-1}N\left(N^TV^{-1}N\right)^{-1}N^T\Big)V^{-1}\Big)R.$$

By writing $N = MJ$ at one crucial point in this formula it reduces to

$$A_M A_N = R^T\Big(V^{-1} - V^{-1}\Big(M\,(M^TV^{-1}M)^{-1}M^T + N\,(N^TV^{-1}N)^{-1}N^T - \\ MJ\,(N^TV^{-1}N)^{-1}N^T\Big)V^{-1}\Big)R$$

$$= R^T\left(V^{-1} - V^{-1}M\,(M^TV^{-1}M)^{-1}M^TV^{-1}\right)R$$

$$= A_M.$$

By setting $J = I$, it's also clear that $A_M^2 = A_M$.

You were asked to consider X_M^2 and $(X_N^2 - X_M^2)$, and as both are quadratic functions of $\mathbf{x}$ we can use Craig's theorem (section 4.2) to show that they are independent by calculating $A_M\,(A_N - A_M)$, which vanishes by the calculation of the previous paragraph.

Answer 76. The number of parameters to be fit, k, is one, and the number of parameters whose significance we want to know, r, is also one. We can write down the X_α^2:

$$\sigma^2 X_{n-k}^2 = \sum (x_i - \bar{x})^2$$

$$= ns^2$$

and

$$\sigma^2 X_{n-(k-r)}^2 = \sum (x_i - \mu)^2,$$

and then calculate f:

$$f = (n-1)\frac{n\bar{x}^2 + n\mu^2 - 2n\bar{x}\mu}{ns^2}$$
$$= \frac{(\bar{x}-\mu)^2}{s^2/(n-1)}.$$

We know that f follows an $F_{1,n-1}$-distribution and that a 1-tailed test is appropriate; $F_{1,\alpha}$ is equivalent to t_α^2, so

$$\frac{\bar{x}-\mu}{s/\sqrt{n-1}}$$

follows a t_{n-1}-distribution, and because of the square root the test is now 2-tailed. This is the same test we arrived at directly in an earlier section.

Answer 77. I find that $f = 2.72$, and tables of F tell me that the 95% and 99% values from an $F_{5,40}$-distribution are 2.45 and 3.51 respectively. I can accept the extra parameters at the 95% level, but not at the 99% level.

You might note that the quoted values of χ^2 don't give much faith in the model, as either justifies its rejection at quite a high significance level. If the model is reasonable the errors may have been underestimated; if they are really Gaussian this won't affect the F-test, but I'd be suspicious.

Answer 78. Substituting for $\hat{\boldsymbol{\theta}}$ we see that the claimed result is equivalent to the identity

$$\mathbf{y}^2 \equiv \mathbf{y}^T\mathbf{y} = \mathbf{y}^T(I - M(M^TM)^{-1}M^T)\mathbf{y} + \mathbf{y}^TM(M^TM)^{-1}M^T\mathbf{y}$$
$$\equiv n\hat{\sigma}^2 + n\tau^2.$$

If $\boldsymbol{\theta} = \mathbf{0}$, $\mathbf{y} = \boldsymbol{\epsilon}$ and this becomes

$$\mathbf{y}^2 = \boldsymbol{\epsilon}^T(I - M(M^TM)^{-1}M^T)\boldsymbol{\epsilon} + \boldsymbol{\epsilon}^TM(M^TM)^{-1}M^T\boldsymbol{\epsilon}$$
$$\equiv \boldsymbol{\epsilon}^TA\boldsymbol{\epsilon} + \boldsymbol{\epsilon}^TB\boldsymbol{\epsilon}.$$

It is easily seen that $AB = 0$, so by Craig's theorem $\hat{\sigma}$ and τ are independent.

This separation of the residuals into two parts is closely related to the standard technique known as the analysis of variance, ANOVA, a large topic which this book does not discuss.[3,4]

Answer 79. Because x and y are independent, $\langle z\rangle = \langle x\rangle\langle y\rangle = \mu_x\mu_y$, and

$$\begin{aligned}\sigma_z^2 &= \langle x^2y^2\rangle - (\mu_x\mu_y)^2\\ &= (\sigma_x^2+\mu_x^2)(\sigma_y^2+\mu_y^2) - \mu_x^2\mu_y^2\\ &= \sigma_x^2\sigma_y^2 + \sigma_y^2\mu_x^2 + \sigma_x^2\mu_y^2,\end{aligned}$$

so

$$\begin{aligned}r_{xz} &= \frac{\Big\langle (xy-\mu_x\mu_y)(x-\mu_x)\Big\rangle}{\sigma_x\sigma_z}\\ &= \frac{\langle x^2y\rangle - \langle xy\rangle\mu_x}{\sigma_x\sigma_z}\\ &= \frac{\sigma_x^2\mu_y}{\sigma_x\sigma_z}\\ &= \frac{\mu_y/\sigma_y}{1+(\mu_x/\sigma_x)^2+\left(\mu_y/\sigma_y\right)^2}.\end{aligned}$$

For example, if $\mu_x = \sigma_x$ and $\mu_y = \sigma_y$, $r_{xz} = 0.577$. This isn't just an academic calculation, as it is quite common to see data presented where the two axes are far from independent.

Answer 80. The estimator $\hat{\theta}$ is a linear function of the errors, so it too is Gaussian. Problem 68 tells us how to find the needed estimator s^2, and your solution to problem 69 shows that this s^2 is independent of $\hat{\theta} - \theta$. Continuing to collect previous results, problem 66 proves that s^2 is a χ^2_{n-k} variable, and using the usual formula for $\hat{\theta}$'s covariance $W(\hat{\theta}) = \sigma^2(M^{\mathsf{T}}M)^{-1}$ we indeed see that

$$t^2 = \frac{(\hat{\theta}_i - \theta_i)^2}{s^2(M^{\mathsf{T}}M)^{-1}_{ii}}$$

is a t^2_{n-k} variable. For the case discussed in section 11.2, $k = 2$, and if we want to test $\theta_2 = 0$, this expression is readily reduced to

$$(n-2)\frac{r^2}{1-r^2},$$

which is in agreement with the previous result.

Answer 81. Taking the proffered hint,

$$V = \sum_{i<j} h_{ij}(j-i) + \sum_{i>j} h_{ij} - \sum_{i>j} h_{ij}$$

$$= \sum_{ij} j h_{ij} - \sum_{i<j} i h_{ij} - \sum_{i>j} h_{ij}$$

$$= \sum_{j} j\left(n - X_j\right) - \sum_{i<j} i$$

(as $\sum_i h_{ij} = n - X_j$ and $h_{ij} + h_{ji} = 1$ if $i \neq j$)

$$= n\sum_{j} j - \sum_{j} jX_j - \sum_{j} j\,(n-j)$$

$$= \sum_{j} j^2 - \sum_{j} jX_j$$

as claimed. Remembering that the X_i are just the first n integers in some strange order, and that the Y_i are just the first n integers, we can calculate the variance of X_i and the value of r_s:

$$\begin{aligned} V(X) &= \frac{\sum X_i^2}{n} - \left(\frac{\sum X_i}{n}\right)^2 \\ &= \frac{(n+1)(2n+1)}{6} - \left(\frac{n+1}{2}\right)^2 \\ &= \frac{n^2-1}{12} \end{aligned}$$

and

$$\begin{aligned} r_s &= \frac{1}{V(X)}\left(\frac{\sum iX_i}{n} - \bar{X}^2\right) \\ &= \frac{1}{V(X)}\left(\frac{\sum iX_i - \sum X_i^2}{n} + \frac{\sum X_i^2}{n} - \bar{X}^2\right) \\ &= 1 - \frac{12V}{n(n^2-1)}. \end{aligned}$$

Answer 82. It is straightforward to write the program. The sample size and number of simulations should, of course, be input parameters.

Running 10000 simulations I estimate that the 97.5th percentile is 0.642. Its variance is given by

$$\frac{p(1-p)}{nf(0.642)^2} = \frac{0.975 \times 0.025}{nf(0.642)^2}.$$

I estimate from my simulations that $f(0.624)$ is about 0.2, so the variance is about $0.6/n$. To obtain an accuracy of 1% I need to run about 15000 simulations. (I could have estimated this in advance by using values from the asymptotic theory. For a Gaussian, the 97.5% percentile is 1.96σ from the mean, let's call it 2σ, so its variance is $0.975 \times 0.025 \times 2\pi e^2/n = 22.6/n$. To obtain an error of 1% we would need the standard deviation to be 0.02, which corresponds to $n = 55000$, which is within a factor of 4 of the correct answer.) The asymptotic theory predicts that

$$\left(\frac{(n-2)r_s^2}{1-r_s^2}\right)^{1/2}$$

follows a t_{n-2}-distribution. The 2-tailed 95% confidence region (which is what we want) is $|t| \leq 2.306$, which corresponds to a value of $r_s = 0.632$ — which is not too different from our numerical value.

Answer 83. If there were very many points in each bin then we would lose hardly any degrees of freedom in estimating the s parameters from the data, and X^2 would still follow a χ^2_{k-1} distribution. You can imagine splitting the dataset in two, estimating the parameters from one half, and conducting the test on the other.

On the other hand, it should be clear that we can't lose more degrees of freedom by estimating from the unbinned data than we lost by using the binned numbers. Taking these two thoughts together, the claimed result should seem plausible.

Answer 84. The d_i would first be positive, then negative, then positive again.

Answer 85. Making the transformation $z = (x - t_1)/t_2$ shows that

$$y_i = \int_{-\infty}^{z_i} f(z; 0, 1)\, dz,$$

whence the usual arguments about cumulative distributions, as rehearsed in problem 19, show that y_i's p.d.f. g is

$$g(y_i) = p(z_i)/f(z_i; 0, 1),$$

where $p(z)$ is the probability of getting a value z; i.e., z's density function.

If f is Gaussian then $f(z_i; 0, 1) = 1/\sqrt{2\pi}\exp(-z_i^2/2)$, but what about $p(z_i) \equiv p((x_i - \bar{x})/\sigma)$? We can let $\sigma = 1$ without loss of generality,

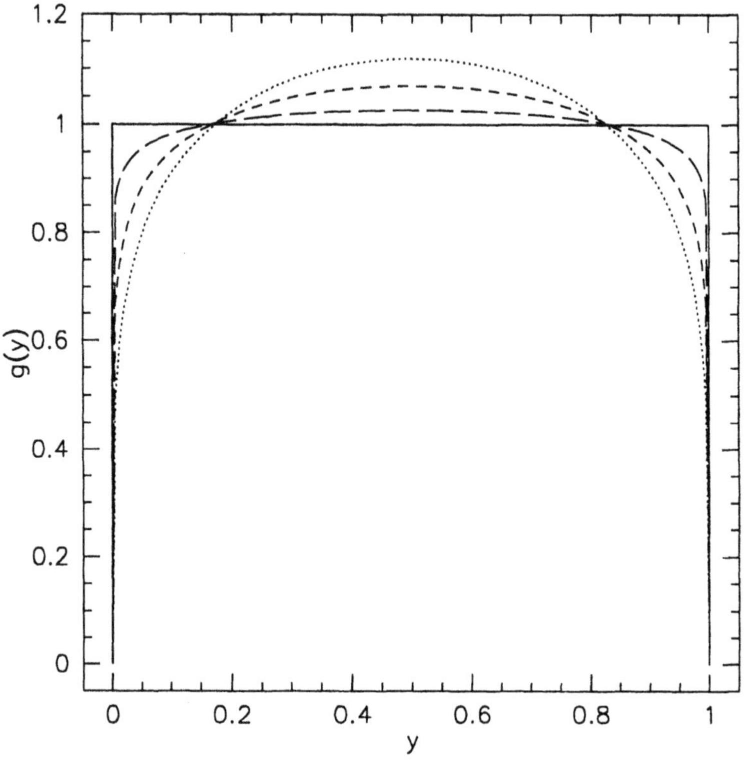

The p.d.f. $g(y)$ for $n = 5$, 8, 20, and ∞.

and write this as $p((n-1)x_i/n - \sum_{j\neq i} x_j/n)$. All the terms are now independent, so z_i's characteristic function is

$$\begin{aligned}\phi_{z_i}(t) &= \phi_x((n-1)t/n)\,\phi_x^{n-1}(-t/n)\\ &= e^{i(n-1)\mu t/n+(n-1)(-i\mu t/n)}e^{-t^2(n-1)^2/2n^2+(n-1)(-t^2/2n^2)}\\ &= e^{-t^2(n-1)/2n}.\end{aligned}$$

This is an $N(0,(n-1)/n)$ distribution and we can write down g:

$$g(y_i) = \sqrt{\frac{n}{n-1}}\,e^{-z_i^2/2(n-1)},$$

where

$$y_i = \frac{1}{\sqrt{2\pi}}\int_{-\infty}^{z_i} e^{-z^2/2}\,dz.$$

Note that these results do *not* depend on the unknown population mean μ. The accompanying figure shows $g(y)$ for a number of values of n.

An almost identical analysis can be used if σ must also be estimated from the data; in this case $p(z_i)$ of course follows a t-distribution.

Answer 86. We know that the distribution of D_n is independent of the mean and standard deviation of the sample, so I chose to use $N(0, 1)$ variates for simplicity.

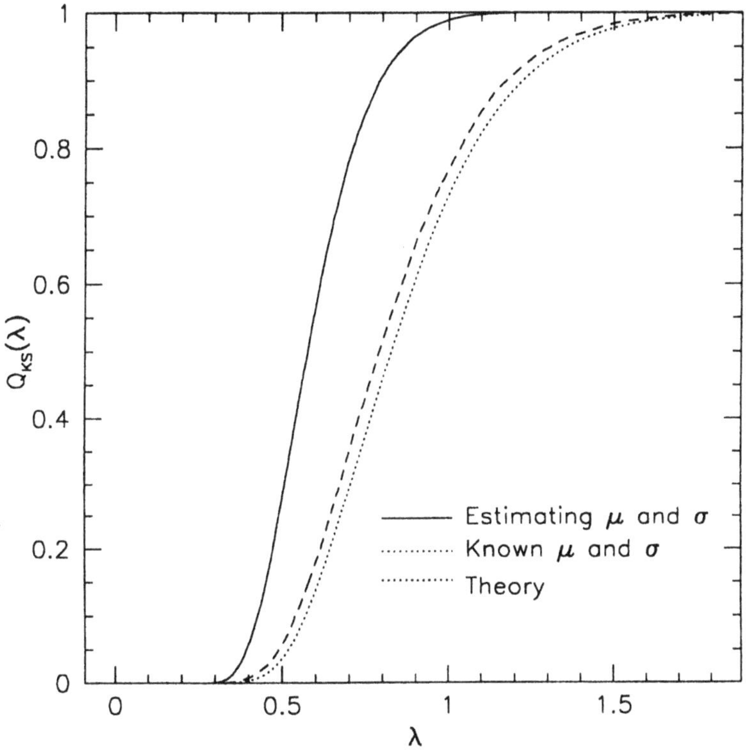

The cumulative distribution of $\lambda = D_n\sqrt{n}$ from 10000 Monte Carlo simulations with $n = 20$.

The 95% point is about 0.865, at which point the value of the p.d.f is about 0.45 (I have not provided a graph of the p.d.f.). The uncertainty in the 95% point is (applying the theory developed in a problem just after discussing the variance of the median)

$$\frac{0.975 \times 0.025}{n(0.45)^2} = \frac{0.35^2}{n}.$$

To achieve an accuracy of 1% we only needed to run about 1200 simulations. The value of λ that we get when we adopt the population values of μ and σ (i.e., 0 and 1) is 1.400; the asymptotic theory predicts 1.354. We would have made a serious error if we had not worried about the effects of estimating parameters, but the asymptotic theory is not too bad, even for such a small sample.

Answer 87. Glancing at problem 36 you will remember that you found the expectation value of $(\bar{x} - \mu)^2$ in a population of size N. Identifying $n_x \equiv n$, $n \equiv N$, $\mu \equiv \bar{z}$, and $\sigma \equiv s_z$, you showed that

$$\left\langle (\bar{x} - \bar{z})^2 \right\rangle = \frac{n - n_x}{n_x(n-1)} s_z^2,$$

so

$$\begin{aligned} \langle w \rangle &= \left\langle \frac{n_x}{n_y s_z^2} (\bar{x} - \bar{z})^2 \right\rangle \\ &= \frac{1}{n-1} \end{aligned}$$

as claimed.

Answer 88. It is easy enough to carry out a two-sample KS test on the two sets of salaries (see the accompanying figure); the value of $D_n \sqrt{n_g n_y / (n_g + n_y)}$ is 1.423.

If the asymptotic result given in the text is applicable, we would expect values at least this large to occur by chance only 3.5% of the time, so we are able to reject the hypothesis that the distributions are identical at the 96.5% level. If we use a χ^2 test we are forced to bin the data; using bin boundaries at every integer and half integer, I find a value of $\chi^2 = 8.95$ for 6 degrees of freedom. A value this large would occur by chance 17.6% of the time, so we can only reject the hypothesis that the distributions are identical at the 82.4% level. How about the variances? If the parent populations are both Gaussian with variance σ^2, then

$$\frac{\sum (g_i - \bar{g})^2}{\sigma^2 (n_g - 1)}$$

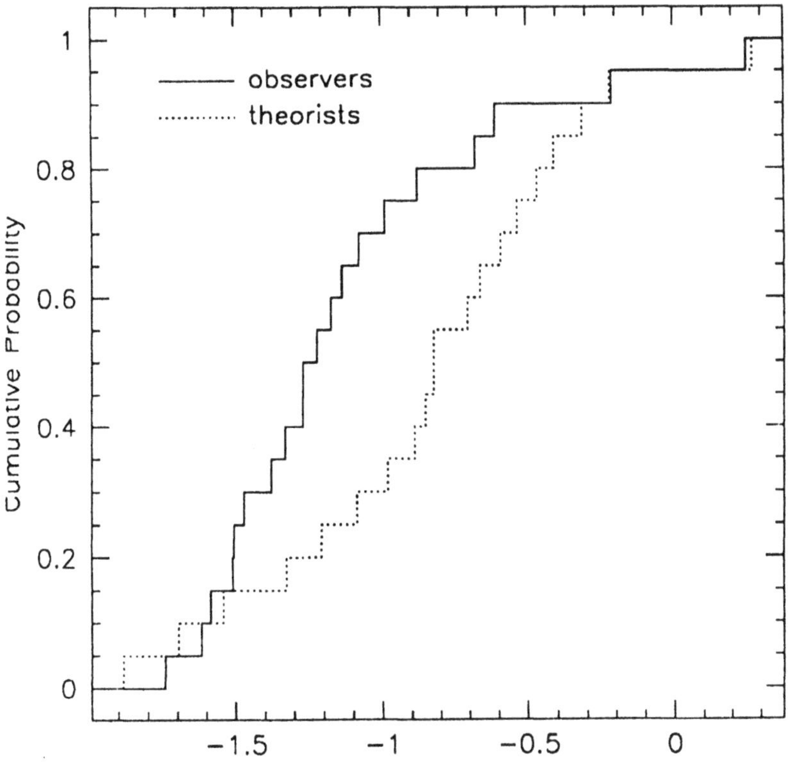

The cumulative probabilities of observers' and theorists' salaries. The significance of the difference is estimated in the text.

and

$$\frac{\sum(y_i - \bar{y})^2}{\sigma^2(n_y - 1)}$$

are respectively distributed as χ^2_ν/ν with $n_g - 1$ and $n_y - 1$ degrees of freedom, i.e., their ratio is distributed as the variance ratio (F) distribution (this is the ratio of unbiased estimates for σ_g^2 and σ_y^2). Carrying out the calculation, we find that $F_{19,19} = 0.948$, a value that would be exceeded by chance 90.9% of the time; we conclude that we cannot reject the hypothesis that the two variances are equal. In passing, I should remind you that this F ratio test is an example of a *non*-robust test based upon variances.

We can now ask whether their means differ significantly; to do this we can either use a Wilcoxon test or a t-test. The value of the Wilcoxon

U-statistic is -2.029, and if we assume that the sample is large enough to use the normal approximation to the distribution of U we see that this value is significant at the 96.5% level. We know that

$$\frac{\bar{x} - \bar{y}}{\sqrt{\frac{n_x s_x^2 + n_y s_y^2}{n_x + n_y - 2}\left(\frac{1}{n_x} + \frac{1}{n_y}\right)}}$$

follows a t-distribution with $n_x + n_y - 2$ degrees of freedom, so a quick calculation shows that $t = -1.745$ with 38 degrees of freedom; a value so different from 0 occurs by chance only 8.7% of the time, so we can say that the means differ at the 91.3% level.

Answer 89. This is an example calling for a KS test. We first estimate the mean and standard deviation of the best Gaussian. We can either use ML estimators, or else the usual unbiased modifications; let us choose that latter, which give $\bar{\epsilon} = 0.308$ and $s = 3.467$. The resulting value of $D_n\sqrt{n}$ is 0.483 — but what is its significance? If we hadn't estimated the parameters of the fit from the data we could use the results in this book to deduce that we'd get a value at least this large 97.4% of the time, a number that might seem unbelievably good (the data are real, by the way). Fortunately there are a couple of problems with this analysis. First, n isn't really all that large, and the asymptotic result is better for nearly-rejected hypotheses than for ones that are easily accepted. Of course, we could calculate (or look up) the true distribution of D_n to rectify this. Second, we estimated the mean and variance from the data, and, as discussed in section 14.4, this invalidates the theoretical distribution of D_n. It is found that the critical values of D_n are about 30% smaller if we estimate the mean and variance from the data rather than somehow knowing the population values.

We can find the value of t for the extra measurement; it's $t = (10.016 - 0.308)/3.467 = 2.8$. For 28 degrees of freedom the probability of finding a value of t so different from 0 is 0.92%, so we can reject the hypothesis at the 99.08% level. Note that if we had foolishly used a normal test rather than a t-test we'd have found a probability of 99.48%, which is quite different (it's the difference from 100% that counts, 0.52% as opposed to 0.92%).

Answer 90. The Pearson correlation coefficient $r_{xy} = 0.4414$. The Spearman correlation coefficient $r_s = 0.4387$, and n is large enough that

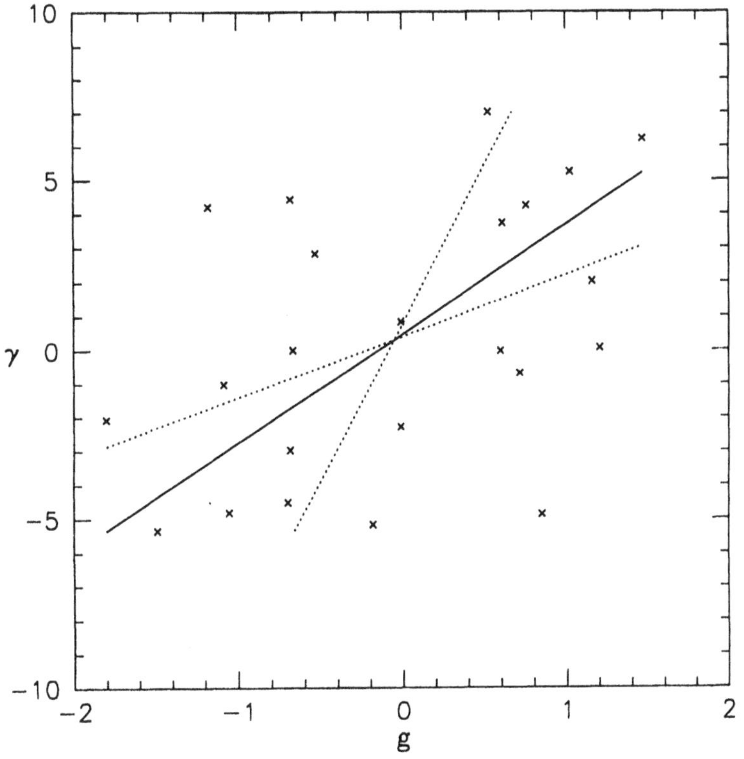

The least-squares fits to the $g : y$ data resulting from the assumptions that all the errors are in g; that they are all in y; and that $\sigma_y = 5\sigma_g$.

we should be able to use the approximation that the distribution of

$$\left(\frac{(n-2)r_s^2}{(1-r_s^2)}\right)^{1/2}$$

follows the t-distribution with $n - 2$ degrees of freedom. The value of t is 2.237 for 21 degrees of freedom; a value so different from 0 would arise by chance only 3.6% of the time (i.e., this is the 2-sided confidence limit). The three different least-squares lines are

$$y = \begin{cases} 1.810g + 0.397 & \text{all errors in } y, \\ 9.292g + 0.769 & \text{all errors in } g, . \\ 3.228g + 0.467 & \sigma_y = 5\sigma_g. \end{cases}$$

These are illustrated in the accompanying figure. The line with errors in both g and y will always lie between the lines giving a pure regression of g on y or y on g.

References

1: *K&S* Ex. 15.6 2: *K&S* 29.22 3: *K&S* 35 4: $K\&S_V$ 29

Symbol Index

Note that Greek symbols are indexed as if they were written out, so that "ρ" is indexed as "rho" and "Θ" as "Theta."

Problem Index

Index

When an index entry is merely a cross-reference, the page number given is that of a likely source of information, so you may not always have to lookup the cross-referenced item.

Note that Greek symbols are indexed as if they were written out, so that "χ" is indexed as "chi."

Ingram Content Group UK Ltd.
Milton Keynes UK
UKHW020805190523
421985UK00004B/108